T0135979

Synthesis Of Branched α- and β-Amino Acids

Using C-Nucleophile Additions to Imines and Nitrones

Von der Fakultät Chemie der Universität Stuttgart
zur Erlangung der Würde eines
Doktors der Naturwissenschaften
(Dr. rer. nat.)
genehmigte Abhandlung

vorgelegt von
Alevtina Baskakova
aus Emwa/Russland

Hauptberichter: Prof. Dr. V. Jäger
Mitberichter: Prof. Dr. B. Plietker
Tag der mündlichen Prüfung: 16.09.2009

Institut für Organische Chemie
der Universität Stuttgart
2009

Bibliografische Information der Deutschen Nationalbibliothek

Die Deutsche Nationalbibliothek verzeichnet diese Publikation in der
Deutschen Nationalbibliografie; detaillierte bibliografische Daten sind
im Internet über http://dnb.d-nb.de abrufbar.

ISBN 978-3-8325-2326-8

Logos Verlag Berlin GmbH
Comeniushof, Gubener Str. 47,
10243 Berlin
Tel.: +49 (0)30 42 85 10 90
Fax: +49 (0)30 42 85 10 92
INTERNET: http://www.logos-verlag.de

Part of this work has been presented in:

Publications

Frey, W.; <u>Baskakova, A.</u>; Jäger, V.

Crystal structure of (3R/S)-3-(9-amino-9-fluorenyl)-azetidin-4-one-2-spiro-9'-fluorene, $C_{28}H_{20}N_2O \cdot 1/8H_2O$

Z. Krist. NCS **2009**, 224, 24-26.

Frey, W.; <u>Baskakova, A.</u>; Jäger, V.

Crystal structure of (2S,3S)-3-(1-adamantyl)-3-[N-(Z)-benzylidene-N-oxyamino]-1,2-isopropylidene-1,2-propanediol, $C_{23}H_{31}NO_3$

Z. Krist. NCS **2009**, 224, 27-28.

Oral presentation

<u>Baskakova, A.</u>; Jäger, V. „On the Way to α- and β-Amino Acids with Bulky Substituents" N, O-Heterocycles and More – 2nd BBS (Bratislava, Berlin, Stuttgart) Symposium on Organic Chemistry, Stuttgart, Germany, April 12 – 15 2007, Book of Abstracts, p. 23.

<u>Baskakova, A.</u>; Jäger, V. „Synthesis of Branched α- and β-Amino Acids Using C-Nucleophile Additions to Imines and Nitrones" N, O-Heterocycles and More – 3nd BBS (Bratislava, Berlin, Stuttgart) Symposium on Organic Chemistry, Berlin, Germany, April 3 – 4 2009, Book of Abstracts, p. 20.

Poster

<u>Baskakova, A.</u>; Jäger, V. "Attempts at the Synthesis of Branched α- and β-Amino Acids" 8. Iminium Salz Tagung, Bartholomä, Germany, September 11 – 13 2007, Book of Abstracts, p. 139.

Table of Contents

Preliminary remarks and abbreviations

Figures, equations, literature citations, schemes, and tables were numbered consecutively.

All compounds prepared during this work and cited in the Experimental Part were consequently numbered **1, 2, 3** etc. and were assembled in the Table of Structures at the end of this work. Some preparations yield diastereomeric mixtures; the diastereomers were assigned as **a** and **b**.

Starting from Chapter 1, all the other formulas and structures are consequently labeled in bold capital letters, i. e. **A, B,..., Z, AA, AB** etc.

Journal abbreviations are given according to the recommendations of the Royal Society of Chemistry (RSC).[1]

List of abbreviations

Ac	acetyl	lit.	literature
Ad	1-adamantyl	*m*	*meta*
Ar	aryl substituent	M	molarity
AIBN	α,α'-azobis(isobutyronitrile)	Me	methyl
Boc	*tert*-butoxycarbonyl	min.	minute(s)
Bn	benzyl	m. p.	melting point
Bu	butyl	MPLC	Medium-Pressure Liquid
c	concentration		Chromatography
CAN	ceric ammonium nitrate	MS	mass spectrometry
conc.	concentrated	NMO	*N*-methylmorpholine-*N*-oxide
COSY	Correlated Spectroscopy	NMR	Nuclear Magnetic Resonance
d	day(s)	N	normality
DEPT	Distortionless Enhancement by	*n*	*normal*
	Polarization Transfer	*o*	*ortho*
DMF	*N,N'*-dimethylformamide	*p*	*para*
DMSO	dimethyl sulfoxide	Ph	phenyl
d.r.	diastereomeric ratio	PG	protecting group
EDAC	1-ethyl-3-(3-	PMB	*p*-methoxybenzyl
	dimethylaminopropyl)	PMP	*p*-methoxyphenyl
	carbodiimide hydrochloride	Pr	propyl
ee	enantiomeric excess	Py	pyridine
Eq(s).	equation(s)	R	(organic) residue
Equiv.	Equivalent(s)	r. t.	room temperature
Et	ethyl	Tos	*p*-toluenesulfonyl
h	hour(s)	*t*	tertiary
HOBT	1-Hydroxybenzotriazole	TFA	trifluoroacetic acid
HPLC	High-Pressure Liquid	THF	tetrahydrofuran
	Chromatography	TLC	Thin-Layer Chromatography
HRMS	High-Resolution Mass	TMS	trimethylsilyl
	Spectrometry	UV	ultraviolet
Fig.	figure	v	volume
i	iso or *ipso* (^{13}C-NMR)	Z	benzyloxycarbonyl
IR	Infrared Spectroscopy		
LAH	lithium aluminium hydride		
LDA	lithium diisopropylamide		
LiHMDS	lithium bis(trimethylsilyl) amide		

Abstract

Artificial peptides containing α- and/or β-amino acids with bulky substituents, may possess some unusual properties, for example, decreased conformational mobility or formation of stable secondary structures. Therefore, the preparation of such unnatural amino acids is of high interest.

In this Thesis, a new route to enantiomerically pure 1-adamantylglycine **D** was developed. The crucial step was the highly stereoselective addition of adamantylmagnesium bromide to the C=N bond of 2,3-*O*-cyclohexylidene-D-glyceraldehyde *N*-benzylnitrone effected in the presence of Lewis acid. The diol moiety served as a masked carboxy group. 1-Adamantylglycine was synthesized in 28 % yield over 6 steps (from the protected D-glyceraldehyde *N*-benzylnitrone).

The routes to 9-aminofluorenylacetic acid **E** and β-amino-β,β-diphenylpropionic acid **F** consist in the addition of allyl Grignard reagents to imines with various protecting/activating groups, followed by oxidative transformation of the allyl group. *N*-Boc-protected 9-aminofluorenylacetic acid **E** was synthesized in 41 % yield over 6 steps (from fluorenone). The methyl ester of β-amino-β,β-diphenylpropionic acid (**F**) in form of its *N*-p-methoxybenzyl derivative was prepared in 6 % yield over 8 steps starting from benzophenone (the yield was not optimized).

Some procedures for the preparation of 3,3-disubstituted 4-azetidinones **46** and **54**, direct precursors of the corresponding β-amino acids, have been investigated.

The best results were obtained by means of the Staudinger reaction. The four-step procedure starting from fluorenone and benzophenone led to the β-lactams **46** and **54** with yields of 39 and 19 %, respectively.

Ausführliche Zusammenfassung und Ausblick

Die besonderen Eigenschaften von α- und β-Aminosäuren mit sperrigen Substituenten sowie deren zahlreiche Anwendungsmöglichkeiten in der chemischen und medizinischen Forschung erwecken ein erhebliches Interesse für asymmetrische Synthesen solcher Verbindungen. Im Rahmen dieser Dissertation sollten 1-Adamantylglycin **D** (eine α-Aminosäure) und die räumlich gehinderten β-Aminosäuren **E–I** synthetisiert werden.

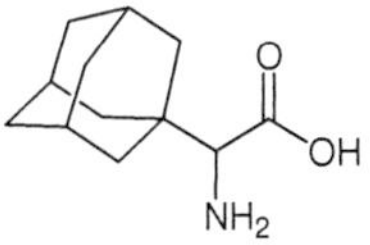

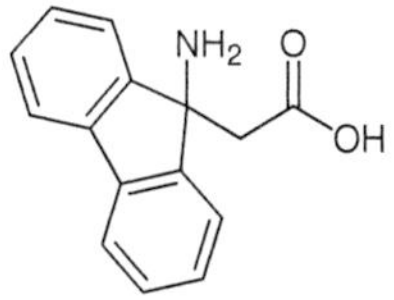

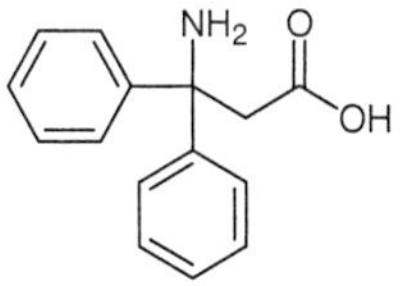

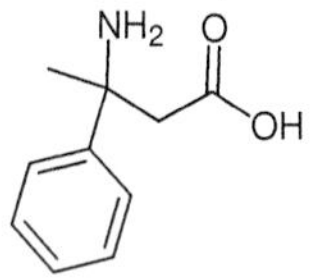

Als Synthesestrategien für die Herstellung solcher α- und β-Aminosäuren wurden die folgenden vorgeschlagen:

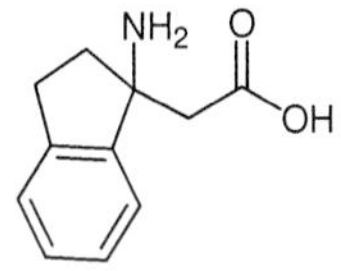

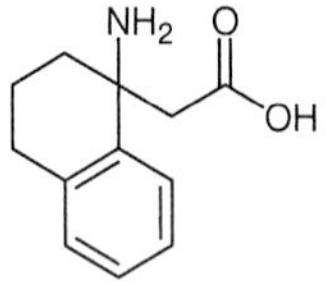

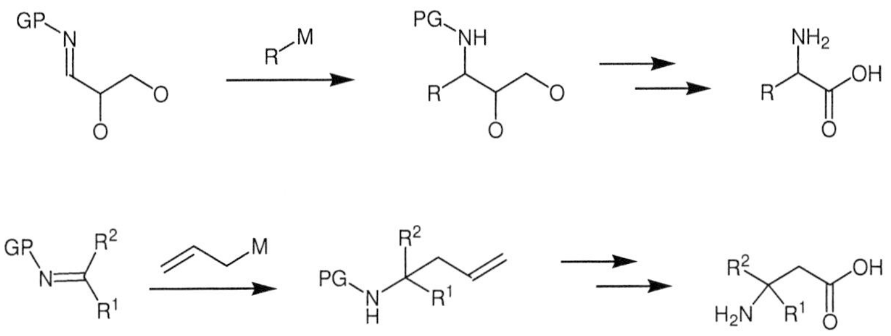

R, R^1, R^2 = Alkyl/Arylrest; M = MgBr, MgCl oder Li; PG = Schutzgruppe

Diese Herstellungsverfahren beinhalten die Grignard-Addition an Imine bzw. Nitrone. Sie können einerseits den Zugang zu einer strukturellen Vielfalt derartiger Aminosäuren gewährleisten und andererseits erlauben, die Stereoselektivität der nucleophilen Additionen zu kontrollieren.

A) Synthese von 1-Adamantylglycin D

In dieser Arbeit wurde eine Reihe von Reaktionen mit Phenyl-, *tert*-Butyl- und 1-Adamantyl-Grignard-Reagenzien durchgeführt. Aufgrund der größeren Stabilität der Nitronen gegenüber den strukturell ähnlichen Iminen wurde das 2,3-*O*-Isopropyliden-D-glyceraldehyd-*N*-benzylnitron (**4**) anstelle eines Imins eingesetzt (Gl. 1). Die Ergebnisse sind in Tabelle I zusammengefasst.

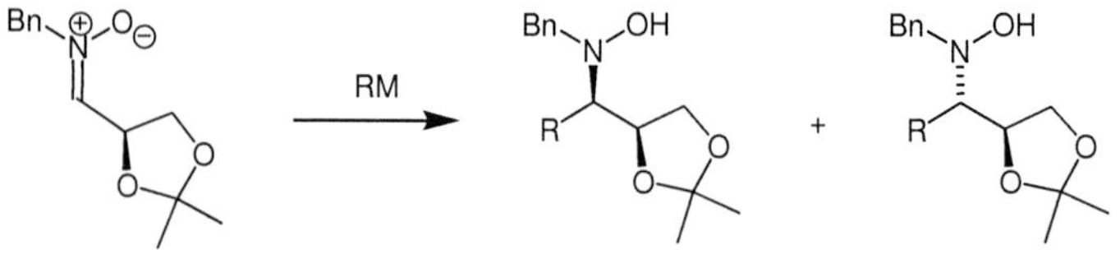

Das Ergebnis der Reaktion von 1-Adamantylmagnesium- und Phenylmagnesiumbromid mit dem Nitron **4** hing von der Gegenwart einer Lewis-Säure ab. Bei der Zugabe von Zinkbromid sank die *syn*-Diastereoselektivität der Addition (Tabelle I, Fälle 1, 2), wobei die Zugabe von Diethylaluminiumchlorid zur umgekehrten Diastereoselektivität führte, so dass ausschließlich das *anti*-Diastereomer gebildet wurde (Tabelle I, Fall 3). Die Addition von Adamantylmagnesiumbromid ohne Präkomplexierung mit einer Lewis-Säure verlief nicht selektiv. Der Zusatz von Diethylaluminiumchlorid führte wiederum zu einem höheren Diastereomerenverhältnis von >95:5.

Die MPLC-Trennung der diastereomeren *N*-Hydroxylamine, die aus dem Nitron **4** erhalten wurden, gelang im Fall R = Ph. Die Konfiguration der *N*-Hydroxylamine mit R = *t*-Bu wurde nicht ermittelt. Die Konfiguration des *N*-Hydroxylamins **7b** (R = Ad) wurde erst nach der Umsetzung zu Adamantanderivaten bestimmt.

Tabelle I. Reaktion von Organomagnesium- und -Organolithiumreagenzien mit 2,3-O-Isopropylidene-D-glyceraldehyd N-Benzylnitron (4)

Fall	RM	Äquivv.	Lewissäure	Zeit	threo : erythro[c]	Ausbeute, Gemisch [%]	Ausbeute der isolierten Diastereomeren [%]
1	PhMgBr	1.5	$ZnBr_2$	6 h	58 : 42	48	–
2	PhMgBr	1.5	keine	6 h	75 : 25	82	33 (**5a**) + 13 (**5b**)
3	PhMgBr	1.5	Et_2AlCl	16 h	<5 : 95	91	70
4	PhLi	3	keine	40 min	85 : 15	73	46 (**5a**) + 9 (**5b**)
5	t-BuLi	1.5	keine	40 min	75 : 25[d]	20	–
6	t-BuMgCl	2	keine	6 h	78 : 22[d]	76	12 (**6a** oder **6b**) 9 (**6a+6b**)
7	AdMgBr	2	keine	17 h	42 : 58	51	–
8	AdMgBr	2	Et_2AlCl	16 h	<5 : 95	42	–

[a] Lösungsmittel: Diethylether, außer Fall 2 (THF). – [b] Temperatur: –60 °C für M = MgBr und –85 °C für M = Li. – [c] Bestimmt aus Intensitäten der ^{13}C NMR Signale. – [d] Die Konfigurationen der Stereoisomeren wurden nicht ermittelt.

Für die Synthese von 1-Adamantylglycin wurde das 2,3-*O*-Cyclohexyliden-D-glyceraldehydnitron **16** verwendet, da die Cyclohexylidenschutzgruppe im Vergleich zu der Isopropyliden-Schutzgruppe gegenüber Säuren stabiler ist (Schema I). Verglichen mit dem Fall 8 (Tabelle I), bei dem das Nitron **4** mit der Isopropyliden-Schutzgruppe verwendet wurde, ergab die Zugabe von Adamantylmagnesiumbromid zum Nitron **16** eine effektive Verbesserung der Ausbeute (61 %). Die *erythro*-Konfiguration von *N*-benzylhydroxylamin **17** wurde erst nach der Umsetzung zum (*S*)-1-Adamantylglycin-Hydrochlorid [(*S*)-**24**] bestimmt.

Das Additionsprodukt **17** wurde in Gegenwart von Pearlman's Katalysator hydriert. Das resultierende Amin wurde in 87 % Ausbeute isoliert und anschließend mit 12 N Salzsäure behandelt. Daraus wurde das Hydrochlorid des 3-Amino-1,2-diols **21** in 82 % Ausbeute erhalten. Dieses Produkt reagierte mit Di-*tert*-butyldicarbonat unter Zusatz von Triethylamin und ergab das entsprechende *N*-geschützte Aminodiol **22** in einer Ausbeute von 79 % (Schema I).

Schema I. Synthese von (S)-1-Adamantylglycinehydrochlorid [(S)-24]

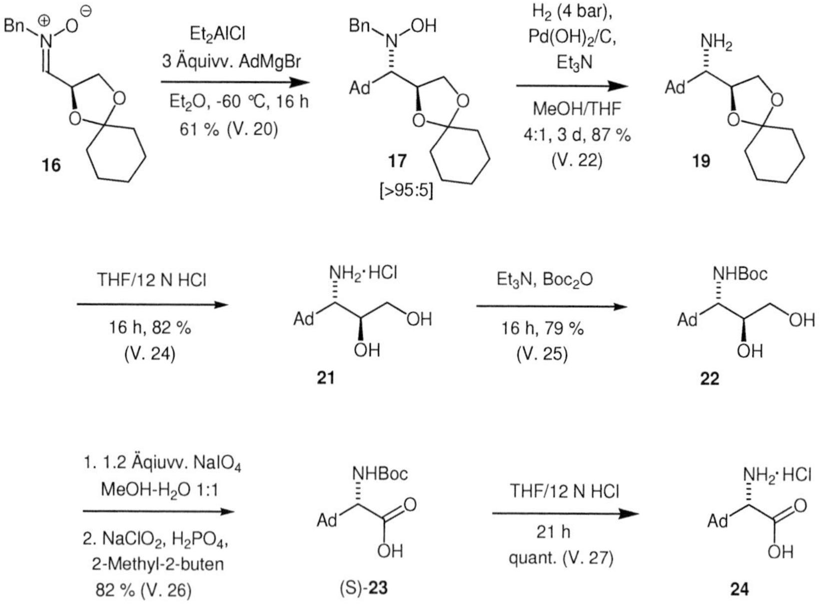

Die Transformation des Diolrestes wurde durch oxidative Spaltung mit Natriumperiodat und anschließende Oxidation mit Natriumchlorit zur Carbonsäure durchgeführt. Dieses Verfahren ergab *N*-Boc-1-Adamantylglycin [(*S*)-**23**] mit guten Ergebnissen (82 % Ausbeute). Die folgende Abspaltung der N-Boc-Schutzgruppe führte zu (S)-1-Adamantylglycin-Hydrochlorid [(S)-24]. Somit wurde die Ziel-Aminosäure in einer Ausbeute von 28 % in 6 Stufen erhalten.

B) Synthese von β-Amino-β,β-diphenylpropionsäure (F)

Benzophenon wurde als einfach zugängliches Ausgangsmaterial gewählt (Schema II). Um das N-geschützte Imin **25** darzustellen, wurden Benzophenon und p-Methoxybenzylamin in Anwesenheit von Zinkbromid unter Rückfluss in Toluol erhitzt. Nach der Addition von Allylmagnesiumbromid an das Imin **25** wurde das Allylderivat **26** in sehr guter Ausbeute erhalten, siehe Schema II.

Schema II. Synthese des N-geschützten β-Amino-β,β-diphenylpropionsäuremethylesters

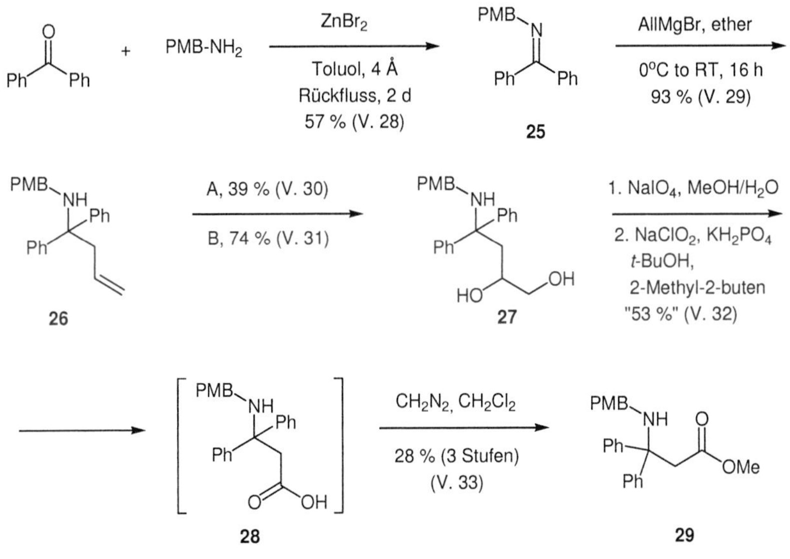

Methode A: 1. 10 Gew.% NMO/H_2O, 2.5 Gew.% OsO_4/t-BuOH, t-BuOH/THF, 6 h. - 2. Na_2SO_3
3. Säulenchromatographie.
Methode B: 1. $K_3Fe(CN)_6$, $K_2OsO_2(OH)_4$, K_2CO_3, t-BuOH/H_2O, 5 d. -
2. Na_2SO_3. - 3. Umkristallisation.

Um das Diol **27** verfügbar zu machen, wurden zwei Methoden getestet (A, B, Schema II). Die Reaktion der Allylverbindung **26** mit Osmiumtetroxid in Anwesenheit von N-Methylmorpholin-N-oxid als Reoxidans lieferte da Diol **27** in geringer Ausbeute von nur

39 %. Die Methode mit Kaliumosmat-Dihydrat/Kaliumhexacyanoferrat System dagegen lieferte das Diol **27** in besserer Ausbeute.

Das Diol **27** wurde zu 3-(4-Methoxybenzylamino)-3,3-diphenylpropansäure (**28**) oxidiert. Die Struktur dieser Verbindung ließ sich nicht belegen: NMR-Spektren konnten nicht gemessen werden, da das Produkt **28** weder in Wasser noch in einem anderen organischen Lösungsmittel sich löste. Um dies zu verbessern, wurde die Säure **28** in den Methylester umgewandelt (Schema II). Abgesehen von der niedrigen Ausbeute von 28 % über 3 Stufen konnte der Methylester **29** so problemlos ausgehend von dem Diol **27** isoliert werden.

Die Verbindung **29** kann als die Zielaminosäure **F** mit geschützten Amino- und Carboxylgruppen betrachtet werden. Demzufolge bleibt die Optimierung der Veresterungsstufe notwendig.

C) Synthese von 9-Aminofluorenylessigsäure (E)

Die Aminosäure **E** wurde gemäß dem bereits vorgestellten Weg synthetisiert: Einführung des Allylrestes durch nucleophile Addition an die C=N-Imin-Bindung mit nachfolgendem oxidativen Abbau der Doppelbindung des Olefins.

Wie die Reaktionen in Schema III zeigen, ergab Fluorenon mit Lithiumhexamethyldisilazid-Lösung das entsprechende *N*-Trimethylsilylimin **38**, welches nicht isoliert, sondern direkt mit etherischer Allylmagnesiumbromid-Lösung umgesetzt wurde. Das 9-Allyl-9-aminfluoren (**39**) wurde in mäßiger Ausbeute (24 %) isoliert.

Verbesserte Ausbeuten wurden erzielt, wenn das Amin **39** nur als Rohprodukt isoliert und ohne weitere Reinigung mit Di-*tert*-butyldicarbonat umgesetzt wurde. Diese "Eintopfmethode" führte zu deutlich verbesserter Ausbeute (61 %) des *N*-Boc-Derivates **40** über 3 Stufen.

Die Doppelbindung des Allylrestes in Verbindung **40** wurde in Gegenwart einer katalytischen Menge von Kaliumosmat-Dihydrat und Kaliumhexacyanoferrat dihydroxyliert (Schema III). Die Umwandlung der 1,2-Diol-Funktion erfolgte durch oxidative Spaltung mit Natriumperiodat und anschließender Oxidation des entsprechenden Aldehyds in Anwesenheit von Natriumchlorit. Die *N*-Boc-geschützte β-Aminosäure **42** wurde schließlich in 6 Stufen mit einer Ausbeute von 41 % erhalten.

Schema III. Herstellung von 9-(N-Boc-Amino)-fluorenylessigsäure (**42**)

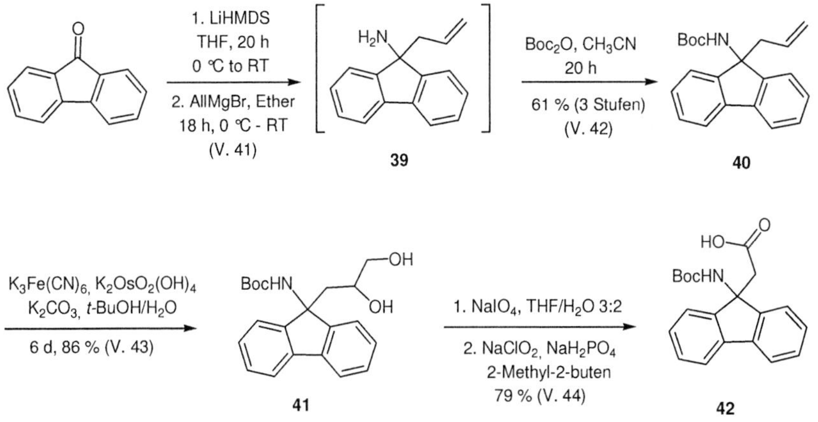

D) Versuche zur Herstellung von Spiro[azetidin-2,9'-[9H]fluoren]-4-on (46)

Mehrere Versuche zur Herstellung des β-Lactams **46** durch Reaktion des N-Trimethylsilylimins **38** mit dem Lithium-Enolat **45** wurden durchgeführt (Schema IV). Das erwartete 2-Azetidinon-Produkt wurde entweder in sehr niedriger Ausbeute oder im Gemisch mit dem Substitutionsprodukt an C-3 des β-Lactam-Rings (**47**) isoliert. Die Resultate dieser Reaktion waren nicht reproduzierbar.

Schema IV. Reaktionen von N-Trimethylsilylfluorenonimin **38** *mit Li-enolat* **45**

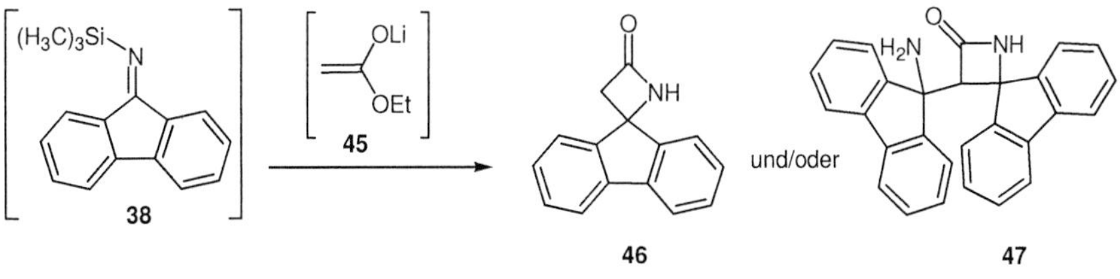

Die N-tert-Butyloxycarbonyl-Schutzgruppe an einem Ketimin wie **38** könnte eine interessante Alternative zu der N-Trimethylsilylgruppe bei Reaktionen mit Lithium-Enolaten sein.

E) Staudinger-Reaktion

Bessere Ergebnisse auf dem Weg zu dem β-Lactam **46** wurden mithilfe der Staudinger-Reaktion erhalten (Schemata V und VI). Die Reaktion von Chloracetylchlorid mit den *N*-*p*-Methoxyphenyliminen **48** und **49** bei Zusatz von Triethylamin ergab die 3-chlorsubstituierten 2-Azetidinone **50** und **52** in guten Ausbeuten.

Das Chloratom in Position 3 wurde durch *Tris*(trimethylsilyl)silan und AIBN als Radikalstarter abgespalten. Die Ausbeuten der *N*-geschützten 2-Azetidinone **51** und **53** waren 86 bzw. 95 %.

Als nächstes wurde die *N*-*p*-Methoxyphenyl-Schutzgruppe entfernt, und ergab, allerdings in mäßigen Ausbeuten, die β-Lactame **46** und **54**. Die Ursache der niedrigen Ausbeuten war, dass die Reaktionsprodukte durch den Überschuss an Cerammoniumnitrat während der Aufarbeitung zersetzt wurden.

Mildere Bedingungen für die Entfernung der *N*-*p*-Methoxyphenyl-Schutzgruppe könnten die Gesamtausbeute der β-Lactame **46** und **54** (43 % bzw. 27 % über 3 Stufen) verbessern.

Diese 2-Azetidinone sind Schlüsselverbindungen auf dem Weg zu den β-Aminosäuren **E** und **F**.

*Schema V. Synthese von β-Lactam **46** durch die Staudinger Reaktion*

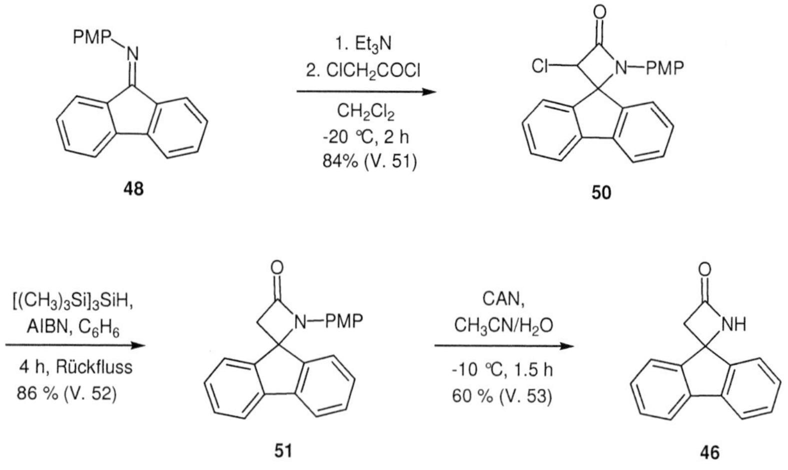

Scheme VI. Synthese des β-Lactams **54** durch Staudinger-Reaktion

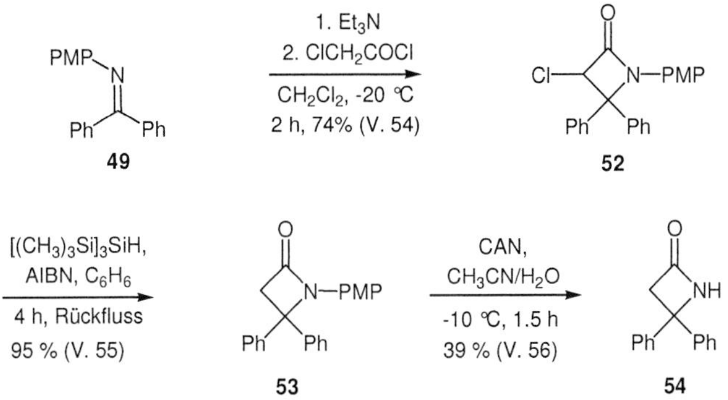

F) Anwendung des Lithium-Enolats von *N,N*-Dimethylacetamid (56)

Abhängig von den Bedingungen führte die Reaktion des *N*-Trimethylsilylfluorenonimins (**38**) mit dem Lithium-Enolat **56** zu einem Gemisch des β-Aminoamids **58** und dem Bernsteinsäurederivat **57** in unterschiedlichen Verhältnissen (Tabelle II). Das Amid **58** ist der Precursor der Zielaminosäure **E**. Es gelang, die optimalen Bedingungen zu finden, bei denen diese Verbindung als Hauptprodukt der Reaktion gebildet wird (siehe Fall 3, Tabelle II). Das *N,N*-Dimethylamid ließ sich weder in 80 % wäßriger Essigsäure noch in der gleichen Lösung nach Zugabe von Salzsäure hydrolysieren. Ein Versuch zur basischen Hydrolyse des β-Aminoamids **58** (in Form seines *N*-Boc-geschützten Derivats) in Gegenwart eines Kalium-*tert*-butanolat/Kaliumhydroxid-Gemisches misslang und die entsprechende Carbonsäure konnte so nicht erhalten werden.

*Tabelle II. Reaktion von the Li-Enolat **56** mit N-Trimethylsilylimin **38***

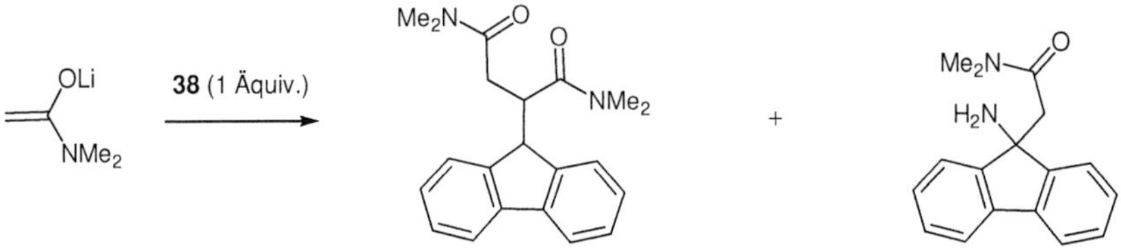

Fall	56 [Äquivv.]	Reaktionsbedingungen	Produkt(e), Ausbeute [%][a]	
			57	58
1	1.5	0 °C, 3 h; RT, 2 h	27	"61"
2	1.2	−30 °C, 16 h	45	"15"
3	1.5	−50 °C, 7 h	traces[b]	77[b]
4	2.5	0 °C, 2 h	"138"[c]	–

[a]Die Ausbeute ist bezogen auf die eingesetzte Menge an Fluorenon. – [b]Neben der Hauptfraktion, zusätzliche Menge an das Amid **58** wurde isoliert (verunreinigt mit Spuren vom Diamid **57**) ("23 %"). – [c]Ein Gemisch mit N,N-Dimethylacetamid.

Ausblick

Der vorgeschlagene Reaktionsweg zu (S)-1-Adamantylglycin beinhaltet den hochdiastereoselektiven Additionsschritt von Adamantylmagnesiumbromid an das Nitron. **4** bzw. **16**. Der Vorteil dieser Route besteht darin, dass die Diastereoselektivität der Addition durch die Wahl der angewendeten Lewissäure beeinflusst werden kann. Das bedeutet, dass beide Enantiomere des substituierten Glycins durch die vorgeschlagene Strategie verfügbar sind. Der neue Weg kann auch für die Herstellung anderer substituierter Glycine mit voluminösen Substituenten wie Phenyl, tert-Butyl und ähnlichen eingesetzt werden.

Die entwickelten Methoden zur Synthese der β-Amino-β,β-diphenylpropion- und 9-Aminofluorenylessigsäure basieren auf der Imin-Addition von Allylmagnesiumbromid. Sie eröffnen den Zugang zu optisch aktiven β-Aminosäuren (**G, H, I** o. ä.). In diesem Fall ist die asymmetrische Induktion (z.B. durch Anwendung der asymmetrischen Katalyse oder eines N-chiralen Auxiliars) bei der nucleophilen Addition notwendig.

1   INTRODUCTION

1.1   Motivation

α-Amino acids are of fundamental biochemical and physiological significance and are essential for life. They serve as starting materials for the synthesis of proteins, which are linear chains of condensed amino acids, and other nitrogen-containing compounds such as purine and pyrimidine bases present in nucleic acids. Amino acids are important for human and animal nutrition, as flavourings, taste enhancers, sweeteners, and others. Both natural and unnatural α-amino acids are also components of many therapeutic agents, agrochemicals, and cosmetics. Because of this wide spectrum of applications in research and industry, a variety of procedures for their extraction from natural sources and for their synthesis in optically pure form has been developed.[2]

Substituted glycines have been utilised as chiral reagents for a variety of synthetic applications. For instance, *tert*-butylglycine (*tert*-leucine) has been the object of intense investigations and many results have been published on its preparations and applications.[3] *N,N*-Dialkyl amides of α-amino acids with a branched side-chain (for example, L-valine, L-isoleucine, L-*tert*-leucine and L-neopentylglycine) are known to take part in highly stereoselective Michael reactions for the construction of quaternary stereocenters.[4]

Nonproteinogenic α,α-dialkylated or α-branched amino acids are conformationally restricted and play a special role in the design of peptides with new interesting properties. The incorporation of such amino acids may enhance the biological activity by decreasing the degree of freedom of the peptide since the rotation around certain single bonds is disabled. These characteristics are interesting for studies of peptide-receptor interactions.[5,6] Moreover, this results in a stabilization of defined secondary structures in small peptides, increased lipophilicity, as well as higher resistance towards both enzymatic and chemical hydrolysis.[7] α,α-Dialkyl α-amino acids themselves are often effective enzyme inhibitors and constitute a series of interesting building blocks for the synthesis of various biologically active compounds.[8]

β-Amino acids, although less frequent in nature than α-amino acids,[9] are also present in peptides, and in free form some of them show interesting pharmacological effects.[10] For instance, emeriamine[11] showed hypoglycemic and antiketogenic activities (in rats, after oral intake); cispentacin[11,12] was shown to be an antifungal antibiotic.

Emeriamine   **Cispentacin**

Functionalized β-amino acids are key components of a variety of bioactive molecules. It was demonstrated that (R)-β-dopa (3,4-dihydroxy-β-phenylalanine), contained in the mushroom *Cortinarius violaceus* as the Fe(III)-catechol complex, gives the blue-violet colour to the fungus. The unsaturated β-amino acid ADDA (3-amino-9-methoxy-2,6,8-trimethyl-10-phenyldeca-4,6-dienoic acid) is incorporated as an active fragment in antibiotic agents, e.g. the microcystins, the nodularins, and motuporin.[11,12]

(R)-β-Dopa   **ADDA**

Some β-amino acids have also been found in naturally occurring peptides with important pharmacological properties. Because of the additional α-methylene group in the "backbone space", β-amino acids represent a class of conformationally more flexible compounds compare to the corresponding α-amino acids. Some peptides, in which an α-amino acid residue was replaced by a β-amino acid at a specific position in the peptide sequence, have shown retention or improvement of the biological activity.[9,12,13] Moreover, certain enzymes destroy such compounds in the human organisms more slowly, because β-peptides form different secondary structures such as β-turns, β-sheets, and helices.[14,15] Representative examples could be cryptophycin (a potent tumour-selective depsipeptide), and jasplakinolide, which is a sponge metabolite with potent insecticidal, antifungal, and antihelminthic properties.[11]

Among β-amino acids, unusual β-branched amino acids are of special interest. Their incorporation into peptide hormones produces highly selective and potent peptide analogues, and at the same time provides new insights into the stereochemical requirements of peptide-receptor interactions.[16]

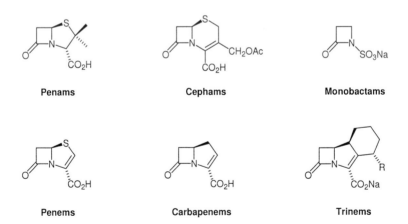

An important class of β-amino acid derivatives are the antibiotic β-lactams.[9,10a,12,13,17] The β-lactam ring can be described as a privileged structure, being part of the structure of several families of antibiotics, principally the penams, cephams, monobactams, penems, carbapenems, and trinems. These antibiotics act by inhibiting the bacterial cell wall synthesis. Inspite of the efficacy of the β-lactam antibiotics, bacteria can become resistant towards commonly used drugs. This requires a permanent development of new, modified structures.

Along with antibacterial properties, some β-lactam derivatives act as anti-inflammatory agents and possess good inhibitory properties against herpes virus.[18]

Discussing properties of α- and β-amino acids and their derivatives, it is necessary to mention β-amino-α-hydroxy and α-amino-β-hydroxy acids since they are key components of a variety of bioactive molecules such as (2*R*,3*S*)-phenylisoserine, which constitutes the side-chain of taxol, or (3*S*,4*S*)-statine, an essential component of pepstatine. Statine and its analoga, (2*R*,3*S*)-norstatine or (4*S*,5*S*)-homostatine,[19,20,21,22] are biologically active structures of natural and unnatural dipeptide isosters showing a strong inhibition of proteases (aspartase, acetylcholinesterase, renin, pepsin, penicillopepsin, cathepsin D).[23] Special cases are presented by the aminopeptidase inhibitors bestatin and amastatin[11] as well as L-hydroxythreonine[24] or polyoxamic acid, being synthetic intermediates in the preparation of β-lactam antibiotics.[25]

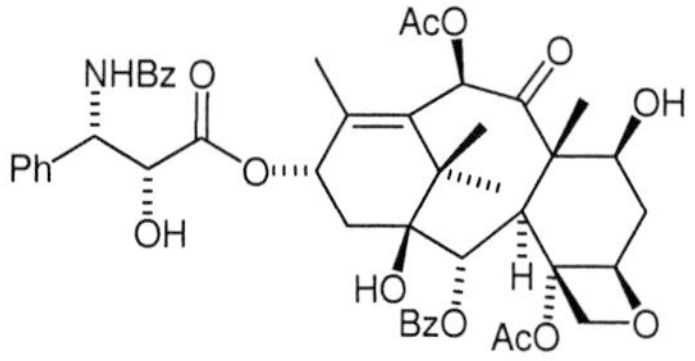

Taxol

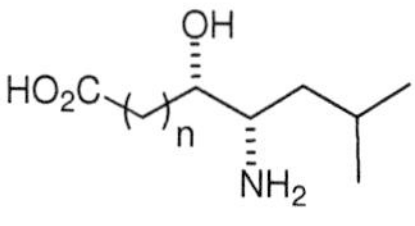

n=1 **Statine**
n=0 **Norstatine**
n=2 **Homostatine**

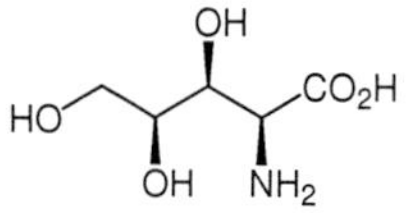

Polyoxamic acid (L-*xylo*)

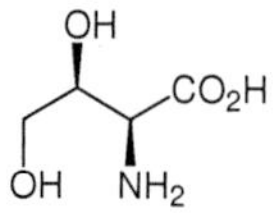

L-Hydroxythreonine

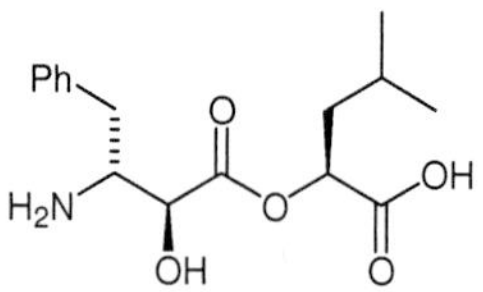

Bestatin

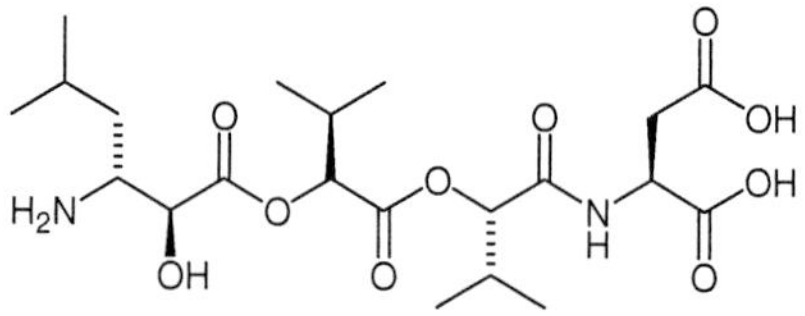

Amastatin

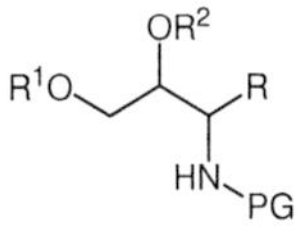

A

R^1 = H, R^2 = benzyl
or R^1– R^2 = isopropylidene
PG = protecting group

Obviously, the 3-amino-1,2-diol fragment is an essential building block of natural and unnatural β-amino-α-hydroxy and α-amino-β-hydroxy acids, where the substituents' configurations affect properties and biological activity. Some methodologies allowing access to the 3-amino-1,2-diol moiety (fragment **A**) were developed and one of them, established in our group, consists in the addition of various organometallic reagents to the C=N

bond[19,20,21,2226] of aldehyde N-benzylimines (eq. 1,2) derived from inexpensive and readily available D-mannitol[27,28,29] or diethyl (-)-2,3-O-benzylidene-L-tartrate[20,30] (eq. 1,2). Organometallics are added to glyceraldimines at room temperature to give amine derivatives as a mixture of both diastereomers (eq. 1,2). The diastereoselectivity of this process was found to depend on the structure of R, the solvent, the metal cation, the type of O-protecting group, and on the presence of a Lewis acid.[31] This dependence upon the reaction conditions allows the control of the stereoselectivity of the addition, and will be considered more closely in Chapter 2.3.1.

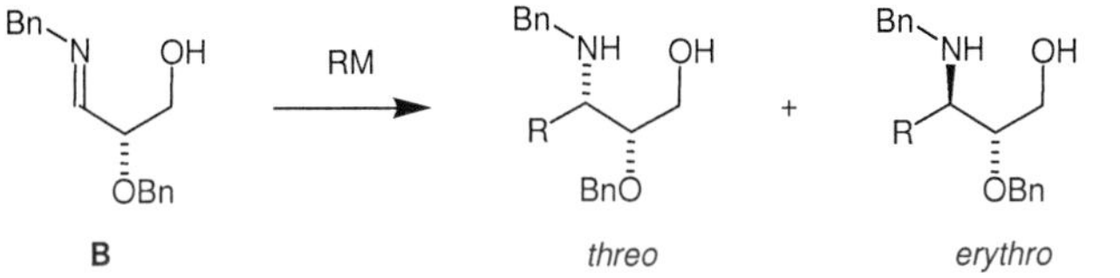

1.2 Definition of the topic

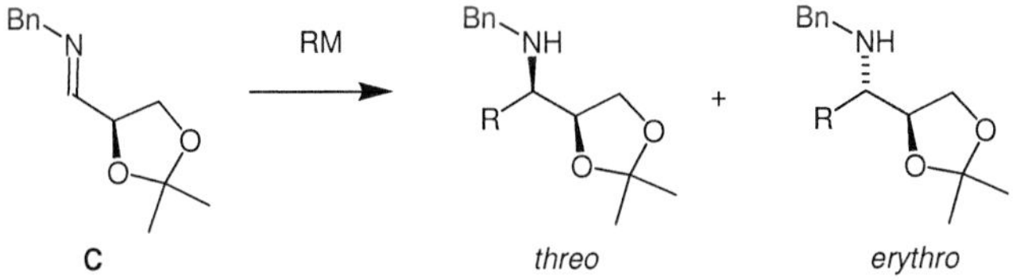

The above-mentioned properties and importance of amino acids for chemical and medicinal research motivated the aim of the present work. It consists in the synthesis of α- and β-amino acids with bulky substituents at the α- and β-positions, respectively. The experience with the additions of organometallics to imines (see eq. 1, 2), gathered in our group, served as a starting point and a key step of this research. Considering this, the following routes to amino acids were proposed:

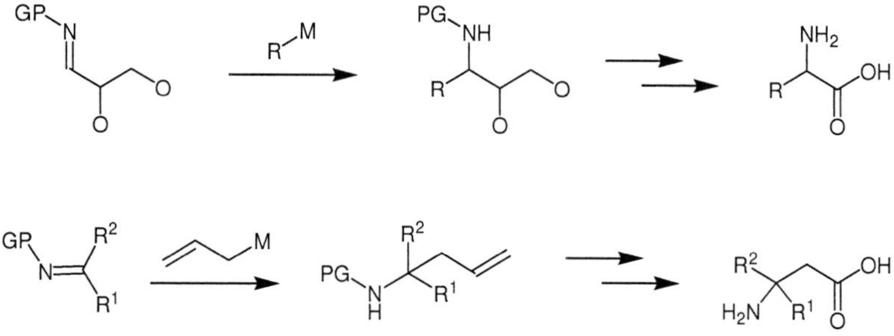

R, R^1, R^2 = alkyl/aryl rests; M = MgBr, MgCl or Li; PG = protecting group

These pathways open an access to a structural variety of amino acids. At the same time, they would allow to control of the stereoselectivity of the imine addition.

To specify the aims of this work, 1-adamantylglycine **D** (α-amino acid) as well as the conformationally constrained β-amino acids **E-I** were chosen as target structures.

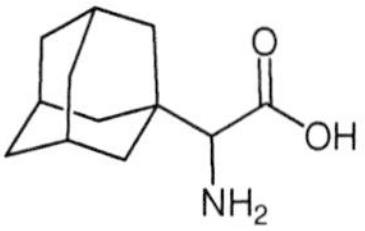

D

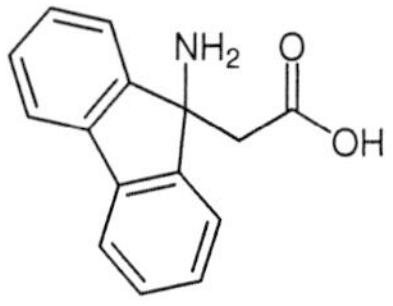

E

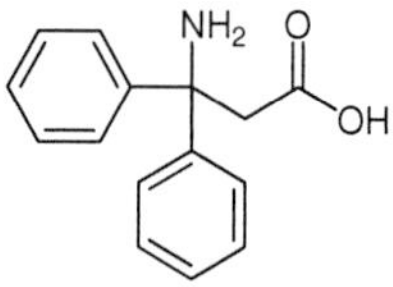

F

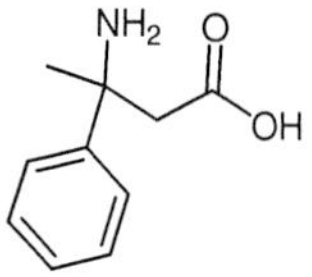

G

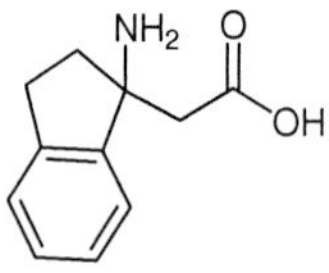

H

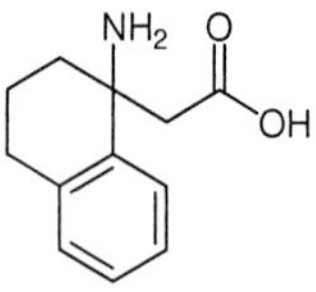

I

2   SYNTHESIS OF (*S*)-1-ADAMANTYLGLYCINE

2.1   Motivation

2.1.1   The adamantane framework as an essential component of biologically active compounds

It has been shown already that substituted glycines with bulky substituents are interesting for the study of peptide-receptor interactions and in asymmetric synthesis. 1-Adamantylglycine is an illustrative example of this type of compounds, and the adamantane framework has been of interest in terms of pharmacological properties for a long time.

The group of Augeri[32] reported on the preparation of saxagliptin® (the synthesis will be discussed below, Scheme 3). Saxagliptin® (as well as galvus®) is currently used for the treatment of type 2 *diabetes mellitus*.[33,34] Memantine® hydrochloride smoothes the consequences of Alzheimer's disease,[35] amantadine® and rimantadine® are used to treat influenza virus A in adults.[36] Tromantadine® is an antiviral medicine used to treat *herpes simplex* virus.[37]

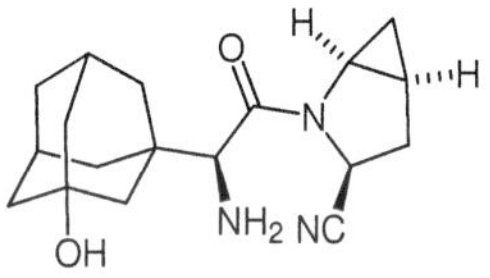

Saxagliptin	Galvus	Memantine

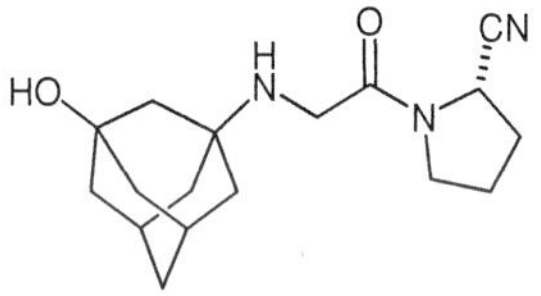

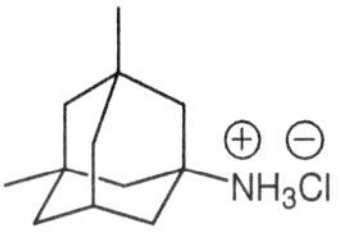

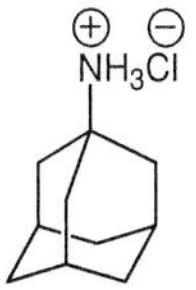

Amantadine	Rimantadine	Tromantadine

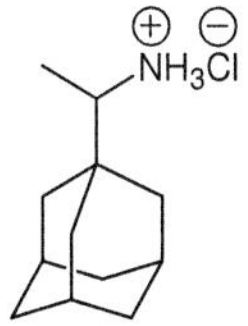

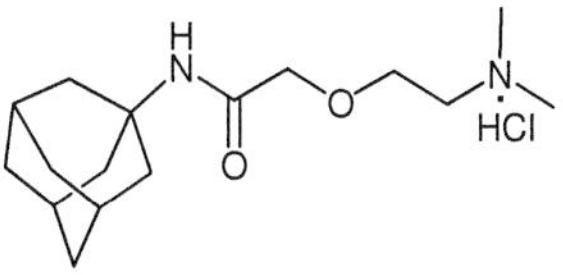

2.1.2   The adamantane skeleton as an active element of a chiral auxiliary

Moreno-Mañas *et al.* used (*R*)-2-(1-adamantyl)-2-aminoethanol as the building block for the synthesis of (*R,R*)-bis(adamantyloxazoline) [(*R,R*)-Adam-Box] (**K**).[38] (*R*)-2-(1-Adamantyl)-2-aminoethanol [(*R*)-**J**] was prepared by enzymatic resolution.[39] The bis(oxazoline) **K** featuring two adamantyl skeletons combined with copper sources makes an excellent catalyst for enantioselective cyclopropanation, Diels-Alder reactions, and allylic oxidations.[38]

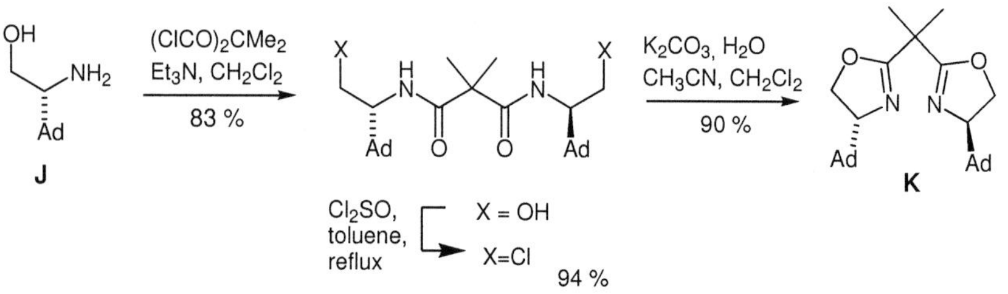

2.2 Known syntheses of 1-adamantylglycines (literature survey)

Both (R) and (S)-enantiomers of 1-adamantylglycine have been prepared earlier by resolution of the racemic mixture.[39] Recently, several publications have appeared dedicated to an alternative synthesis of these compounds. These will be presented in the following Chapters 2.2.1 and 2.2.2.

2.2.1 Assistance of Co[II] complexes of acetoacetic acid derivatives

One of the methods for the synthesis of both enantiomers of 1-adamantylglycine is the alkylation of Co[II] complexes of enantiopure derivatives of acetoacetic acid. This method is suitable for reactions with alkyl halides prone to react through radical-based mechanisms, and substituents such as benzhydryl, 9-fluorenyl and 1-adamantyl were efficiently introduced at the intercarbonyl position.[5,6]

Alkylation of the complex **M** (derived from the β-ketoamide **L** with Evans' oxazolidinone) with 1-bromoadamantane gave, after chromatographic separation, two diastereomeric alkylation products. According to the authors, the low diastereomeric excess is due to the crucial experimental conditions (140 °C). The diastereomer **N** was then converted to (S)-**24** by Schmidt degradation and hydrolysis. The yield of (S)-1-adamantylglycine hydrochloride starting from the β-ketoamide **L** was 20 % over 5 steps[5] (Scheme 1). The latter was obtained from 2,2,6-trimethyl-4H-1,3-dioxin-4-one in 79 % yield as follows:[6]

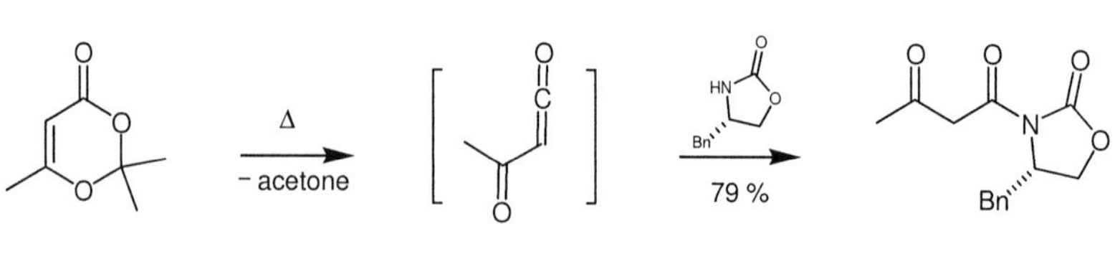

Later, a different approach has been developed.[40] The cobalt(II) complex of ethyl acetoacetate was converted to racemic 1-adamantylglycine (24) (Scheme 2). This was consequently reduced to the corresponding racemic 2-(1-adamantyl)-2-aminoethanol (O) and, after Z-protection of the amino group, exposed to enzymatic resolution in tert-butyl methyl ether at 40 °C, using Pseudomonas capacia lipase (PSL) as biocatalyst and vinyl acetate as acyl donor. The (S)-enantiomer is acylated by the enzyme faster than the (R)-enantiomer and can be separated and converted into the acid by oxidation with sodium metaperiodate in the presence of catalytic amounts of ruthenium(III) chloride. This sequence afforded the N-Z-protected amino acid (R)-Q, which upon hydrogenolysis and further treatment with 7 M HCl afforded (R)-1-adamantylglycine hydrochloride, (R)-24, in 7.9 % after 9 steps (Scheme 2); the specific rotation had the opposite sign compared to that previously described for the (S)-enantiomer.[5]

Scheme 1. Preparation of (S)-1-adamantylglycine hydrochloride[5]

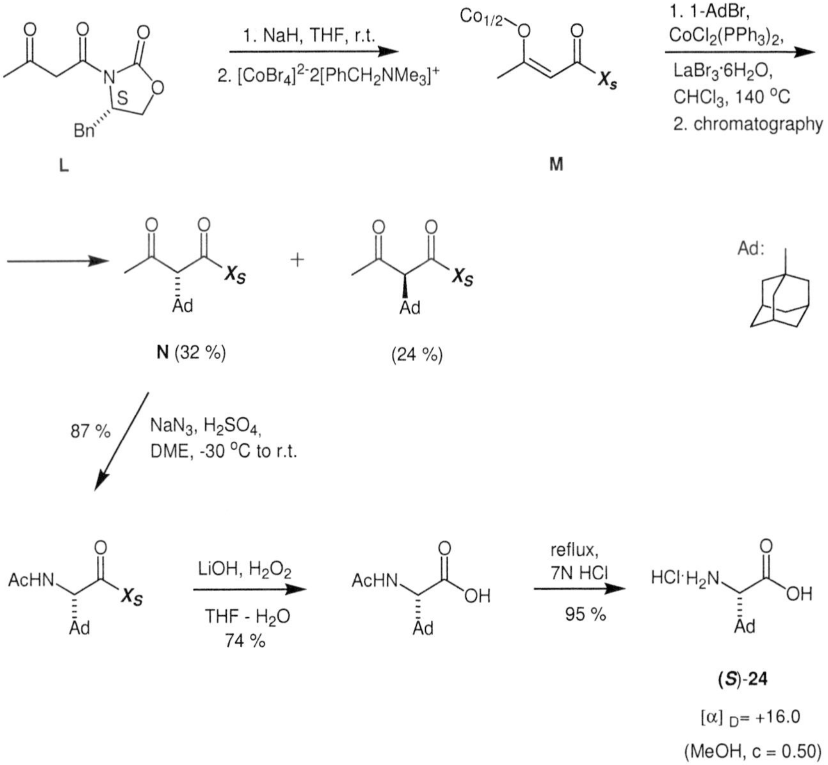

Synthesis of (R)-1-adamantylglycine hydrochloride [40]

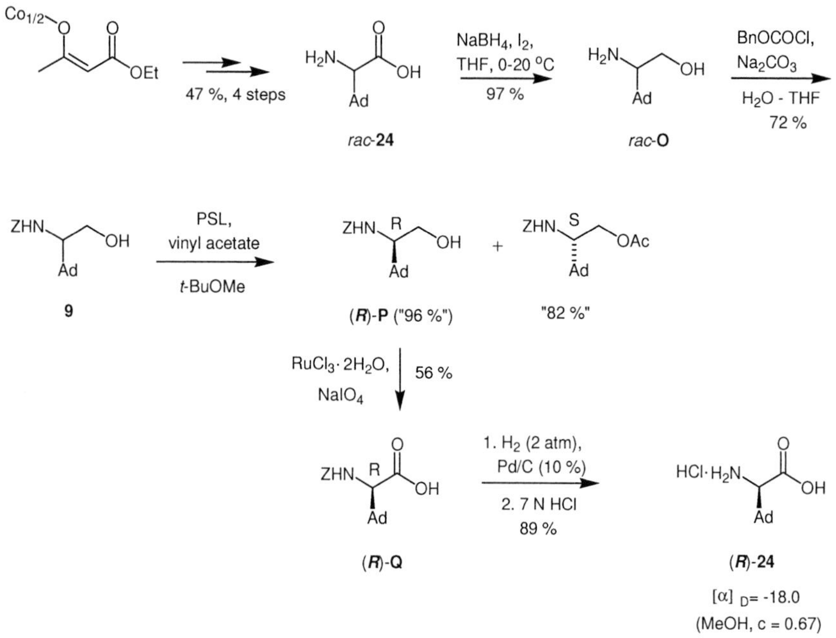

2.2.2 Application of the asymmetric Strecker reaction

The group of Augeri[32] on the way to saxagliptin® developed a fundamentally new approach for the synthesis of the optically active N-protected 1-adamantylglycine. The preparation of (S)-N-Boc-1-adamantylglycine is demonstrated in Scheme 3.

The synthesis employed commercially available adamantane carboxylic acid methyl ester as the starting material. Reduction by lithium aluminium hydride, followed by Swern oxidation, afforded the requisite aldehyde, which was then subjected to asymmetric Strecker conditions – condensation with (R)-(–)-2-phenylglycinol with addition of potassium cyanide – to give the enantiomerically pure R,S diastereomer **R** in 65 % yield. Hydrolysis of the nitrile group gave the acid **S**, followed by hydrogenolysis of the chiral auxiliary, to afford the enantiomerically pure (S)-N-Boc-1-adamantylglycine [(S)-**23**] (43.9 % yield, 6 steps). Hydroxylation of N-Boc adamantylglycine at the bridgehead was accomplished using potassium permanganate in 2 % aqueous potassium hydroxide to give N-Boc-hydroxyadamantylglycine **T** in 51 % yield. Standard acylation conditions were used to couple this to the methanoprolinamide core, furnishing the amide **U** in high yield. This amide was subsequently dehydrated with trifluoroacetic acid anhydride, followed by basic hydrolysis of the resulting trifluoroacetate and deprotection of the amino group. The yield of saxaglyptin® was 16.6 % over 10 steps.

Scheme 2. Synthesis of Saxagliptine[*32]

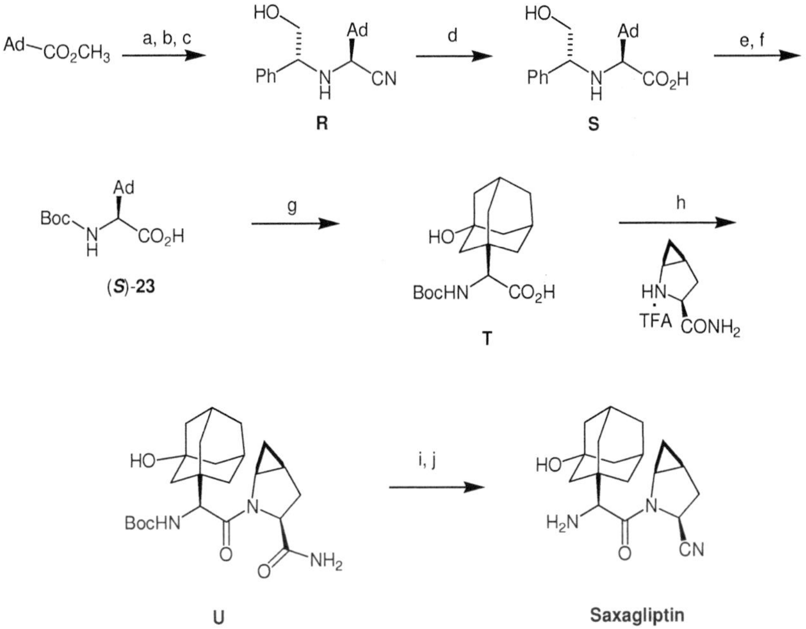

U   Saxagliptin

*Stereodrawings as done by the authors[32]

Reagents and conditions: (a) LAH, THF, 0 °C to RT, 96 %. - (b) (ClCO)$_2$ DMSO, CH$_2$Cl$_2$ -78 °C, 98 %. - (c) (*R*)-2-phenylglycinol, NaHSO$_3$, KCN, 65 %. - (d) 12 N HCl, AcOH, 80 °C, 16 h, 78 %. - (e) 20 % Pd(OH)$_2$, 50 psi, H$_2$, MeOH/AcOH 5:1. - (f) (Boc)$_2$O, K$_2$CO$_3$, DMF, 92 % (2 steps). - (g) KMnO$_4$, 2 % aq. KOH, 60-90 °C, 60 min, 51 %. - (h) EDAC, HOBT, DMF, 85 %. - (i) TFAA, Py, THF, 0 °C to RT, then 10 % aq. K$_2$CO$_3$/MeOH, 92 %. - (j) TFA, CH$_2$Cl$_2$, RT, 95 %.

2.3   Preparation of substituted glycines *via* 3-amino-1,2-diols

2.3.1   Synthesis of aminodiols by addition of metalorganic reagents to *N*-protected imines (literature survey)

In the present work another, alternative way to 1-adamantylglycine was envisaged. It was based on the research performed in our group for several years. The interest in this research was awakened by the fact that the 3-amino-1,2-diol fragment A is an essential part of such type of compounds like natural and unnatural β-amino-α-hydroxy and α-amino-β-hydroxy acids as well as α-amino acids. The method under consideration consists in the addition of various organometallic reagents to the C=N bond of N-benzylaldimines[19,20,21,22] (Eqs. 1,2).

B   threo   erythro   (1)

C   threo   erythro   (2)

Organometallic reagents are added to glyceraldimines at room temperature to give amine derivatives as a mixture of both diastereomers (eqs. 1, 2). The diastereoselectivity of this process was shown to depend on the structure of R, the solvent, the metal cation, the type of O-protecting group, and also on the presence of a Lewis acid.[31] This dependence upon the reaction conditions allows the control of the stereoselectivity.

In 1964 Yoshimura and Sato[41] were the first who had tried the addition of phenyllithium and phenylmagnesium bromide to the N-phenyl- and N-benzyl-substituted 2,3-O-isopropylidene-D-glyceraldimines (eq. 3). The authors could show the dependence of the stereoselectivity on the metal cation used (Table 1). During the addition of phenyllithium, the preferred formation of the syn-diastereomers was observed (entries 1, 3). At the same time, the addition of phenylmagnesium bromide led to the formation of the erythro (anti)-diastereomers (entries 2, 4).

1. 3-4 eq. PhM, Et$_2$O
0 °C, 1 h, reflux, 10 h

2. 6 N HCl, 100 °C, 3 h
Yields: not stated[41]

threo (syn)   erythro (anti)   (3)

Table 1. Additions of organomagnesium- and -lithium reagents to N-protected 2,3-O-isopropylidene-D-glyceraldimines by Yoshimura und Sato[41]

Entry	R	PhM	*threo : erythro*[a]
1	Ph	PhLi	*74 : 26*
2	Ph	PhMgBr	*9 : 91*
3	Bn	PhLi	*63 : 37*
4	Bn	PhMgBr	*22 : 78*

[a] The diastereomeric ratios were determined by the values of specific rotation after conversion to phenylglycine.

The addition of organomagnesium compounds to the *N*-benzyl-2,3-*O*-isopropyliden-D-glyceraldimine **V** was then widely investigated in the group of Cativiela[42,43,44,45] and some others.[46,47,48] The diastereoselectivity of the reactions is difficult to predict. Some results are summarized in Table 2.

Table 2.[a] *Addition of organomagnesium compounds to the N-benzyl-2,3-O-isopropyliden-D-glyceraldimine by Cativiela et. al. V*[45]

Entry	RMgX	Yield [%]	T [°C]	*syn : anti*
1	CH₃MgBr	36	-30	*55 : 45*
2	PhCH₂MgCl	64	-20	*≥ 98 : 2*
3	CH₂=CHCH₂MgBr	76	-30	*70 : 30*
4	PhMgBr	69	0	*≤ 2 : 98*
5	CH₂=CHMgBr	81	0	*≤ 2 : 98*

[a] All reactions were carried out in diethyl ether.

The cerium- and copper-organic reagents are useful alternative to Grignard reagents. Terashima *et al.*[49] have tried the addition of cyclohexylmethylmagnesium bromide to the *N*-benzylamine of 4-*O*-benzyl-2,3-*O*-isopropylidene-D-threose (**W**). No addition has been observed. However, when cyclohexylmagnesium bromide was first treated with cerium(III)

chloride, the addition proceeded smoothly in a highly stereoselective manner, giving rise to
the amine **X** (Scheme 4) as the sole product. On the other hand, treatment of **W** with
cyclohexylmethylcopper (I) in the presence of $BF_3 \cdot Et_2O$ gave the *erythro* product **Y** only.

*Scheme 3. Addition of cerium- and copper-organic reagents to an optically active threose
imine* **W**[49]

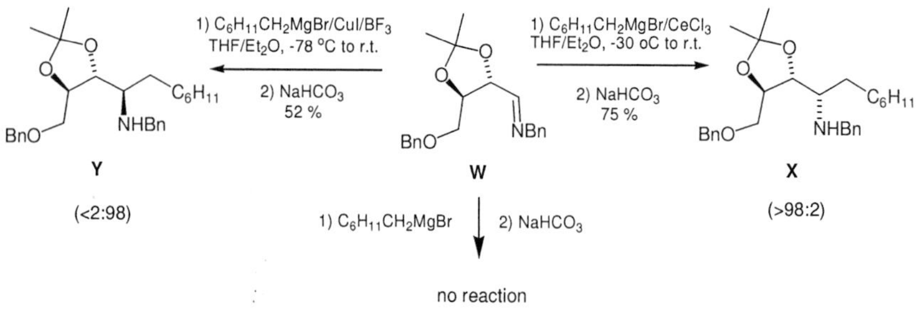

The addition of several organometallic compounds to *N,O*-benzyl-protected glyceraldimines
under various conditions has also been investigated in our group.[19,20,21,22,26,30] Some results
are presented in Table 3[19,20,21,22] (eq. 1). Preferred formation of the *threo (syn)* diastereomer
was normally observed, except when a Lewis acid ($CeCl_3$) was added. The addition of
methyl-, *tert*-butyl-, 1-adamantyl-, allyl- as well as of benzylmagnesium bromide in diethyl
ether gave the products of nonselective addition (entries 1, 7, 8, 11, 12). The sterically
hindered β-substituents isobutyl and cyclohexylmethyl were added with good selectivities
(entries 4, 6). As entries 13 and 14 show, the addition of organometallic compounds using
THF as solvent gives a slower reaction than when ether is employed and the selectivity is
low. The addition of organolithium compounds showed a lack of stereoselectivity (entries 15–
18). $CeCl_3$ as an additive promoted high *erythro*-selectivity (entries 19, 20).

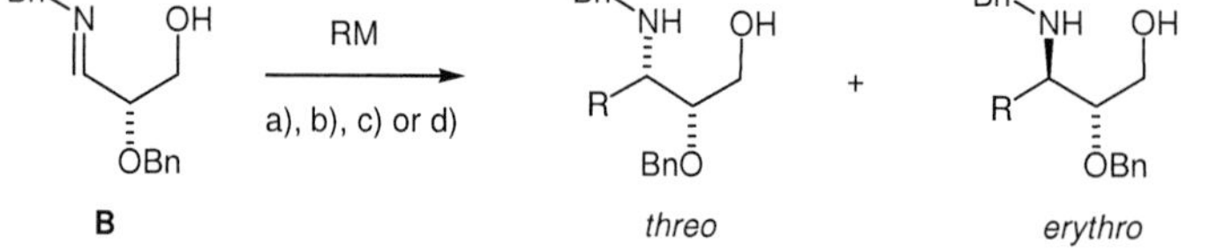

Conditions: a) 4 R'MgX, diethyl ether, 0 °C→ r. t., 12 h – 3 d (entries 1-12); b) 4 R'MgX, THF,
0 °C→ r. t., 3 d c) 2.5 R'Li, diethyl ether, −78 °C→ r. t. (entries 15–18); d) 4 R'MgX/4 $CeCl_3$, 0
°C→ r. t., 12 h (entries 19–21).

Table 3. *Addition of metalorganic reagents to N,2-O-dibenzylglycerinaldimine* **B**[19,20,21,22]

Entry	RM	Yield [%]	*threo:erythro (syn:anti)*
1	MeMgBr	78	*63 : 37*
2	*n*-BuMgBr	73	*73 : 27*
3	*i*-PrMgCl	81	*80 : 20*
4	*i*-BuMgBr	92	*88 : 12*
5	3-MeBuMgBr	89	*90 : 10*
6	CyCH$_2$MgBr	58	*>95 : 5*
7	*t*-BuMgBr	86	*56 : 44*
8	AdMgBr	41	*55 : 45*
9	PhMgBr	61	*76 : 24*
10	VinylMgBr	91	*83 : 17*
11	AllylMgBr	74	*40 : 60*
12	BnMgBr	87	*40 : 60*
13	*n*-BuMgBr	79	*46 : 54*
14	*i*-BuMgBr	93	*47 : 53*
15	MeLi	80	*46 : 54*
16	*n*-BuLi	81	*57 : 43*
17	*t*-BuLi	79	*40 : 60*
18	PhLi	83	*44 : 56*
19	MeMgCl/CeCl$_3$	69	*9 : 91*
20	*i*-BuMgBr/CeCl$_3$	62	*<5 : 95*
21	BnMgCl/CeCl$_3$	70	*76 : 24*

N,2-*O*-Dibenzyl-L-threose imine acetonide **Z** was also subjected to the addition of a series of organometallic reagents, as depicted in the equation 4.[22,26] The products of the additions were hydrolyzed under acidic conditions, without isolation.

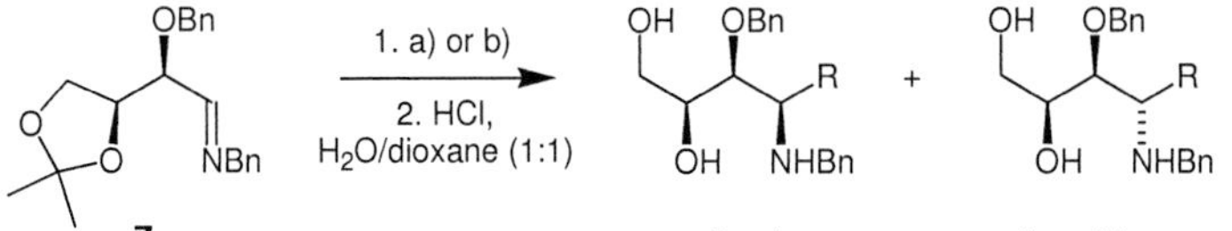

$$\text{(4)}$$

Conditions: a) 3 eq. RLi, diethyl ether, −78 °C, 15 min. b) RMgX, diethyl ether, 0 °C to r. t., 2–5 h; quenching with aqueous NH$_4$Cl.

Table 4. Addition of organometallic reagents RM to the N-benzylthreoseimine Z[22,26]

Entry	RM	Yield[a] [%]	D-xylo/L-arabino[b]
1	MeLi	52	89 : 11
2	n-BuLi	59	>95 : 5
3	t-BuLi	70	>95 : 5
4	Me_3SiCH_2Li	67	>95 : 5
5	PhLi	66	>95 : 5
6	$2\text{-}C_4H_3SLi$	37	>95 : 5
7	VinylMgBr	65	>95 : 5
8	AllylMgBr	75	85 : 15
9	$PhCH_2MgBr$	65	>95 : 5
10	$AnCH_2MgCl$	62	>95 : 5

a) Yield after chromatography on silica gel. b) Diastereomer ratio from ^{1}H and ^{13}C analyses of crude additions products (prior to the acidic hydrolysis).

In contrast to the unpredictable results from Table 3, the additions to the N,2-O-Dibenzyl-L-threose imine **Z** showed excellent 3,4-threo-selectivity in all cases [except entries 1 and 8, where the erythro-adduct (L-arabino-isomer) was also found].

2.3.2 Synthesis of aminodiols by addition of metalorganic reagents to N-protected imines (own results)

The synthesis of substituted glycines via 3-amino-1,2-diol fragment is an important application of the addition of organometallic compounds to N-benzaldimines. This was used as the starting point for this work. In spite of the fact that there are many publications and dissertations[19,20,21,22,26,30] where such additions were quite successful (see discussion above), the present work met with some difficulties. Several attempts to get the 3-amino-1,2-diol moiety by addition of organometallic reagents to N-benzaldimines **B** and **C** (eq. 1, 2) have been carried out, but in all cases the NMR spectra of the crude products were inconclusive; the work-up failed to give a pure compound. The results are summarized in Table 5.

Table 5.[a] Addition of metalorganic reagents to N,2-O-dibenzylglycerinaldimine **B** *and N-benzyl-2,3-O-isopropylideneglyceraldimine* **C**

Imine	Solvent	RM [equiv.]	Yield[b] [%]	Experiment number
B	ether	PhLi, 1.1	70	AB 039[43]
B	ether	PhMgBr, 4	84	AB 033[43]
C	ether	PhLi, 1.1	98	AB 017[19,20]
C	ether	PhMgBr, 2.5	85	AB 019[19,20]
C	THF	PhMgBr, 2.5	65	AB 020[19,20]

[a]Metalorganic reagent was added at 0 °C, then the mixture was stirred at r. t. for 1–3 days. [b]Yield of the crude product.

The reactions like those depicted in equations 1 and 2 gave only a complex mixture of compounds that could not be characterized by NMR spectroscopy. Purification by column chromatography was unsuccessful. This is probably due to the low reactivity of the C=N bond of imines and the tendency of enolizable imine derivatives to undergo deprotonation (eventually followed by reductive dimerisation).[31]

One of the possibilities to avoid these problems is to increase the electrophilicity of the carbon atom of the C=N bond by *N*-alkylation, *N*-oxidation, *N*-acylation, or *N*-sulfonylation, to give more reactive imines, iminium salts, nitrones, *N*-acylimines, *N*-sulfonylmines,[31] or *N*-diphenylphosphinoylamides,[50] respectively:

| *N*-Alkyl/arylimine | *N*-Sulfonylimine | *N*-Acylimine | Nitrone | *N*-Diphenyl-phosphinoylamide |

In this work, nitrones were chosen as suitable substrates because the hydroxylamines obtained by the addition of organolithium or Grignard reagents could easily be deoxygenated by Zn–Cu(OAc)$_2$ in acetic acid[51,52] or catalytic hydrogenation,[53] to give the 3amino-1,2-diols which might then be transformed into α-amino acids. Moreover, nitrones are usually solids which can easily be purified and stored for a long time in contrast to the corresponding imines.

2.3.3   Additions of organometallic reagents to 2,3-O-isopropylidene-D-glyceraldehyde N-benzylnitrone (4) (literature survey)

A series of papers of Dondoni, Merino *et al.* reported on moderate stereoselectivities using the readily available 2,3-O-isopropylidene-D-glyceraldehyde N-benzylnitrone (**4**), when subjected to Grignard additions (Table 6, eq. 5).[51,53,54]

$$R = Ph\ \mathbf{5a} \qquad\qquad R = Ph\ \mathbf{5b}$$
$$R = Et\ \mathbf{AAa} \qquad\qquad R = Et\ \mathbf{AAb}$$
$$threo\ (syn) \qquad\qquad erythro\ (anti)$$

(5)

Table 6. Addition of phenyl- and ethylmagnesium reagents to 2,3-O-isopropylidene-D-glyceraldehyde N-benzylnitrone (4)[51]

Entry[a]	R	Solvent	Lewis acid	threo : erythro[b] (syn : anti)	Yield [%]
1	Ph	THF	none	73 : 27	84
2	Ph	ether	none	65 : 35	78
3	Ph	ether	ZnBr$_2$	78 : 22	86
4	Ph	THF	Et$_2$AlCl	29 : 71	83
5	Ph	ether	Et$_2$AlCl	15 : 85	78
6	Et	THF	none	75 : 25	74
7	Et	ether	ZnBr$_2$	78 : 22	71
8	Et	ether	Et$_2$AlCl	30 : 70	81

[a]The reactions were carried out at −60 °C for 6 h. [b]easured from the intensities of ¹H NMR signals.

In the absence of Lewis acid the *threo (syn)* adduct was the major product of the reaction, with the best results obtained in THF (entries 1, 6). When the reaction was carried out after treatment of the nitrone **4** with 1.1 eq of ZnBr$_2$, the *syn* selectivity increased slightly (compare entries 2 and 3; 6 and 7). In contrast, the use of diethylaluminium chloride as a precomplexing agent led to reversal of the diastereofacial selectivity, thus affording the corresponding *erythro (anti)* isomers as major adducts (entries 4, 5, 8); similar results have been observed in other studies.[53,54,55]

According to this strategy, the diastereoselectivity of the addition can somewhat be controlled towards the required isomer. As a result, 2,3-O-isopropylidene-D-glyceraldehyde N-benzylnitrone (**4**) might be utilized as a convenient precursor in the synthesis of various α-substituted glycines.

This nitrone, as mentioned above, is available from the commercially available D-mannitol through the steps of protection[27] and oxidative cleavage[27] of the diol moiety followed by condensation[56] with N-benzylhydroxylamine[57,58,59] (Scheme 5):

Scheme 4. Preparation of 2,3-O-isopropylidene-D-glyceraldehyde N-benzylnitrone (4)

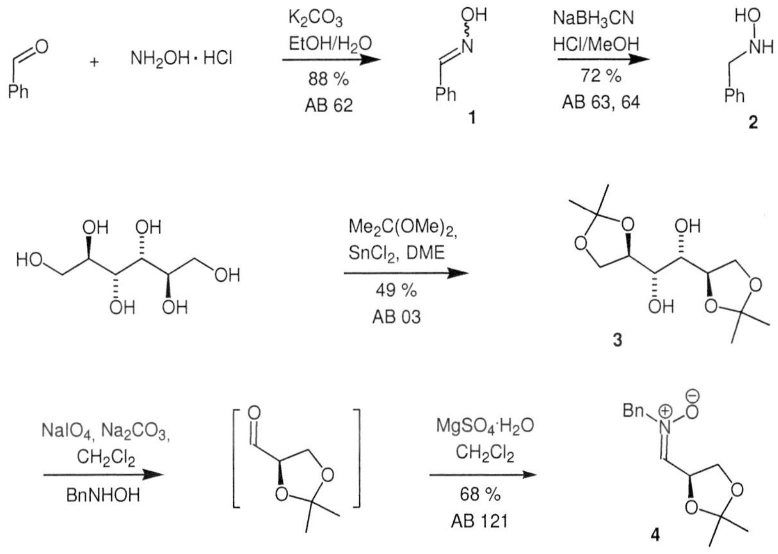

2.3.4 Additions of organometallic reagents to N-benzyl-glyceraldehyde nitrones (own results)

A series of reactions was carried out in this work with phenyl, *tert*-butyl, and 1-adamantyl Grignard reagents in analogy with earlier reports.[51,53,54,55] It was observed that a prolonged reaction time, namely overnight, was necessary to get results comparable to those in the literature. Decreasing the reaction temperature from −40 °C[51,55] to −60 °C led to higher selectivity.[51,55] The results obtained are summarized in Table 7 (eq. 5).

The success of the addition of 1-adamantylmagnesium- and phenylmagnesium bromides to the nitrone **4** was shown to depend on the presence or absence of a Lewis acid. In the case when ZnBr$_2$ was employed, the *syn*-diastereoselectivity of the addition decreased (Table 7, entries 1, 2). In the presence of Et$_2$AlCl the reverse diastereoselectivity was seen, giving exclusively the *anti*-diastereomer (Table 7, entry 3). The addition of adamantylmagnesium bromide without precomplexing with any Lewis acid was non-selective; the presence of Et$_2$AlCl led to an increase of the diastereomeric ratio up to >95:5.

The attempts to separate the diastereomeric *N*-hydroxylamines derived from **4** were successful only in the case of R = Ph, where MPLC separation was carried out using the eluent mixture methylene chloride/isopropanol (97:3 v/v). The *threo* (*syn*) diastereomer **5a** was isolated in yields of 33 and 46 % (entries 2 and 4, respectively), the values of optical rotation were $[\alpha]_D^{20} = -10.8$ (c = 1.00, CHCl$_3$) and $[\alpha]_D^{20} = -16.0$ (c = 1.37, CHCl$_3$); the literature value is $[\alpha]_D^{20} = -6.5$ (c = 1.0, CHCl$_3$).[51,53,55] The *erythro* (*anti*) compound **5b** was isolated as a minor diastereomer in lower yield (13 and 9 %, entries 2 and 4, respectively). Optical rotation values were $[\alpha]_D^{20} = -20.0$ (c = 1.00, CHCl$_3$) and $[\alpha]_D^{20} = -29.8$ (c = 2.03, CHCl$_3$), comparing to $[\alpha]_D^{20} = -17.5$ (c = 1.0, CHCl$_3$) in the literature.[51,53,55] Since literature and recorded values were not satisfactorily close to each other, the NMR data were also used to confirm the configuration of the diastereomers **5a** and **5b**.

The configurations of the *N*-hydroxylamines with R = *t*-Bu were not established. The configuration of the *N*-hydroxylamine **7b** (R = Ad) could be determined after transformations into the adamantane derivatives.

Some reactions were also carried out with the nitrone **AB**, but only starting material was reisolated, though the reaction temperature was increased to room temperature (Table 8).

Table 7. Addition of some organomagnesium and -lithium reagents to 2,3-O-isopropylidene-D-glyceraldehyde N-benzylnitrone (4)

Entry	RM	Equiv.	Lewis acid	Time	threo : erythro[c]	Yield, mixture [%]	Yield of isolated diastereomers [%]	Experim. number
1	PhMgBr	1.5	ZnBr$_2$	6 h	58 : 42 (78 : 22)[51]	48 (86)[51]	–	1
2	PhMgBr	1.5	none	6 h	75 : 25 (73 : 27)[51]	82 (84)[51]	33 (5a) + 13 (5b)	2
3	PhMgBr	1.5	Et$_2$AlCl	16 h (6)[51] h	<5 : 95 (15 : 85)[51]	91 (78)[51]	70	5
4	PhLi	3	none	40 min	85 : 15	73	46 (5a) + 9 (5b)	3
5	t-BuLi	1.5	none	40 min	75 : 25[d]	20	–	6
6	t-BuMgCl	2	none	6 h	78 : 22[d]	76	12 (6a or 6b) 9 (6a+6b)	7
7	AdMgBr	2	none	17 h	42 : 58	51	–	8
8	AdMgBr	2	Et$_2$AlCl	16 h	<5 : 95	42	–	9

[a]Solvent used: diethyl ether, except of entry 2 (THF). – [b]Temperature: –60 °C for M = MgBr and –85 °C for M = Li. – [c]Measured from the intensities of ^{13}C NMR signals. – [d] The configurations of the stereoisomers have not been established.

*Table 8. Addition of some organomagnesium and -lithium reagents to N, 2-O-dibenzyl-L-glyceraldimine N-oxide (**AB**)*

AB

Entry	RM [equiv.]	Lewis acid	Time [h]	T [°C]	Recovered AB [%]	Experiment number
1	PhMgBr (2)	ZnBr$_2$	6	–60	69	AB 51
2	PhLi (1.5)	none	16	–75	79	AB 74
3	PhMgBr (2)	none	50	20	72	AB 78
4	t-BuMgCl (2)	none	7	–60	77	AB 60
5	t-BuMgCl (1.5)	none	18	–65	70	AB 73
6	AdMgBr (3)	ZnBr$_2$	60	0→r. t.	69	AB 96

2.4 Synthesis of (S)-phenylglycine hydrochloride (11·HCl)

The results described above were applied for the preparation of well-known phenylglycine to prove the potential of the method suggested in this work. The product of the addition of phenylmagnesium bromide to the nitrone **4** (Scheme 5) – the N-benzylhydroxylamine **5b** (*anti*-product) – was hydrogenated in the presence of Pd(OH)$_2$/C to remove the N-benzyl and N-hydroxy groups.[52,53] The free amino group was protected *in situ* due to the presence of Boc anhydride in the reaction mixture. This process afforded the N-Boc-protected derivative **8a** in 56 % yield, whereas the same procedure for the *threo(syn)* derivative **8b** gave a better result – 68 % yield. The protected aminodiol **8a** was then subjected to treatment with p-toluene sulfonic acid in methanol as solvent to remove the isopropylidene protecting group.[47] t-Butoxycarbonyl protection, usually very sensitive towards the presence of an acid, stayed intact under these conditions.

The resulting diol **9** has opened the access to N-Boc phenylglycine by oxidation of the 1,2-diol moiety. Final oxidative cleavage in the presence of catalytic amounts of RuCl$_3$ and sodium periodate as a reoxidant in a CH$_3$CN/CCl$_4$/H$_2$O mixture[60] has failed to give the required acid. The transformation was performed by oxidative cleavage with sodium periodate in methanol-water giving the corresponding aldehyde, followed by oxidation to the N-Boc-protected amino acid **10** with sodium chlorite,[61,62] however in moderate yield. Addition of potassium dihydrophosphate to the oxidizing system was important to achieve a basic medium. 2-Methyl-2-butene served as a scavenger to remove sodium hypochlorite and hydrochloric acid that formed during the reaction[63] and that could destroy the starting material or product.

Scheme 5. Preparation of (S)-phenylglycine hydrochloride

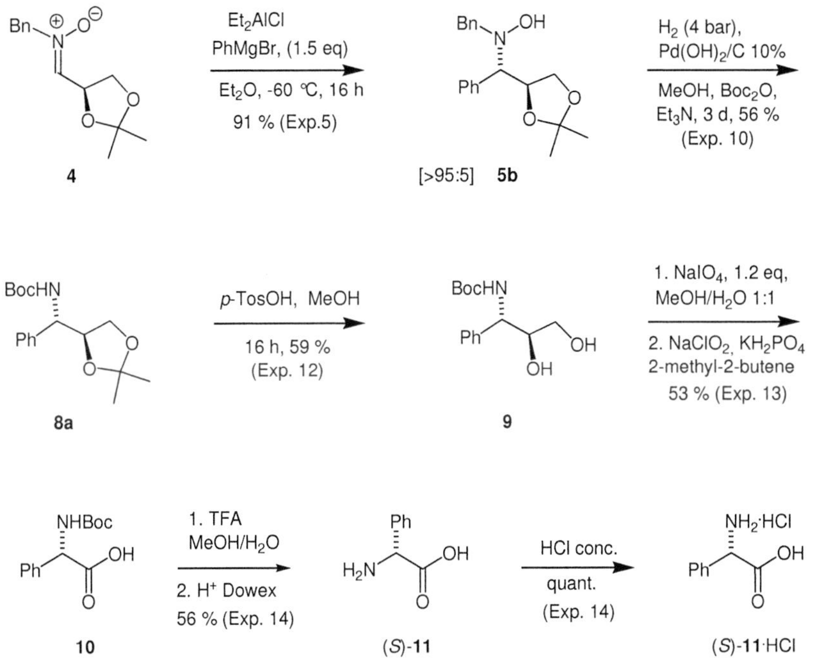

The absolute configuration of the (S)-N-Boc-phenylglycine (**10**) was confirmed by comparison of the value of optical rotation with the literature data.[64] Removal of the N-protecting group was done by means of trifluoroacetic acid and gave, after ion-exchange chromatography, free (S)-phenylglycine in 56 % yield. Treatment with concentrated hydrochloride acid gave (S)-phenylglycine hydrochloride in quantitative yield and 9 % overall yield starting from the nitrone **4**. The ion-exchange chromatographic purification does not appear to be a necessary step, since the N-Boc protected phenylglycine could easily be transformed into the corresponding hydrochloride **11·HCl**. This is discussed below (see Chapter 2.5).

2.4.1 Preparation of (S)-2- tert-butoxycarbonylamino-1-phenylethanol

As illustrated in Scheme 6, the N-Boc aminodiol **9** is a direct precursor for the corresponding protected phenylglycine. On the other hand, it is useful for the synthesis of 2-amino-1-phenylethanol **12**. The diol function of 3-phenyl-1,2-propanol **9** was oxidatively cleaved by sodium periodate, and the aldehyde obtained was reduced with sodium borohydride;[65,66] the yield was 75 % over two steps. 2-Amino-1-phenylethanol **12** after deprotection of the amino group gives 2-phenylglycinol being used as a chiral auxiliary, for example in the synthesis of saxagliptin, as depicted in Scheme 3.[32]

Scheme 6. Synthesis of (S)-2-tert-butoxycarbonylamino-1-phenylethanol[65,66]

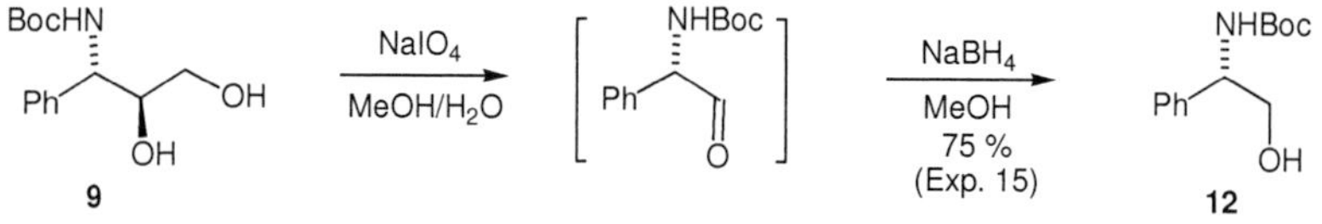

2.5   Preparation of (S)-N-Boc-1-adamantylglycine

2.5.1   Addition of adamantylmagnesium bromide to the 2,3-O-isopropylidene- and 2,3-O-cyclohexylidene-D-glyceraldehyde nitrones

As described above, the addition of adamantylmagnesium bromide to the 2,3-O-isopropylidene-D-glyceraldehyde N-benzylnitrone (**4**, Chapter 2.3.4) was highly diastereoselective and gave the corresponding N-benzylhydroxylamine **7b** in 42–48 % yield (Exp. 9, AB 154) (Scheme 8).

The simultaneous removal of the N-benzyl and N-hydroxy groups was carried out in the presence of Pearlman's catalyst Pd(OH)$_2$/C.[52,53] The amino group released was protected *in situ*, giving the N-Boc derivative **13** in 61 % yield.

The yield of the product **7b** after the addition of adamantylmagnesium bromide was moderate. Probably, in this case the advantage of the isopropylidene protecting group – lability in the presence of acids – turns into a disadvantage: This group can be cleaved even with a weak Lewis acid. In order to avoid this problem, the 2,3-O-cyclohexylidene-D-glyceraldehyde nitrone **16** was used, where the cyclohexylidene protecting group is more stable with respect to acids. In fact, when cyclohexylidene was used instead of isopropylidene, the yield improved to 61 % for the *N*-benzylhydroxylamine **17** (Exp. 16, Scheme 9). The configuration (*erythro*) of **17** was determined later, after conversion into (*S*)-*N*-Boc-1-adamantylglycine.

*Scheme 7. Addition to the 2,3-O-isopropylidene-D-glyceraldehyde nitrone **4** and formation of the nitrone byproduct **14***

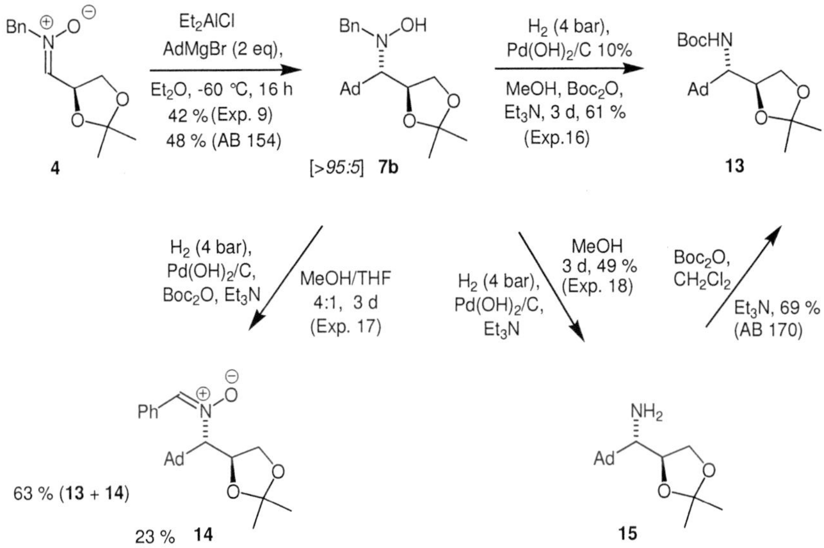

$$63 \% \ (\mathbf{13} + \mathbf{14})$$

23 % **14**

15

The 2,3-*O*-cyclohexylidene-D-glyceraldehyde nitrone **16** was synthesized (Exp. 19) in analogy to the preparation of the corresponding 2,3-*O*-isopropylidene-protected nitrone (**4**, Scheme 5). The structure of **16** was confirmed by X-ray crystal structure analysis (Figure 1).

*Figure 1. X-ray diffraction analysis of the crystal structure of the nitrone **16***

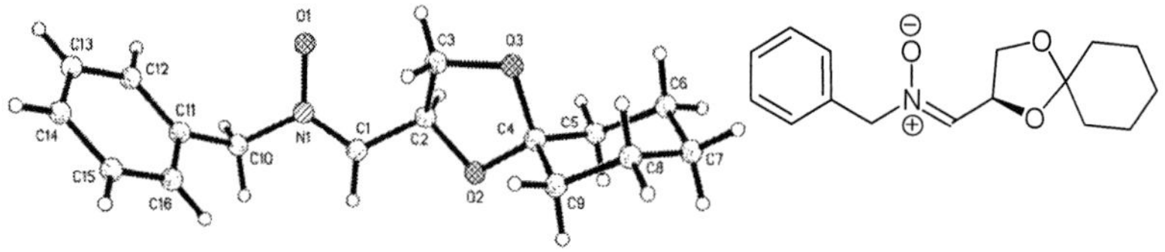

*Scheme 9. Application of 2,3-O-cyclohexylidene-D-glyceraldehyde nitrone **16** to the synthesis of 1-adamantylglycine*

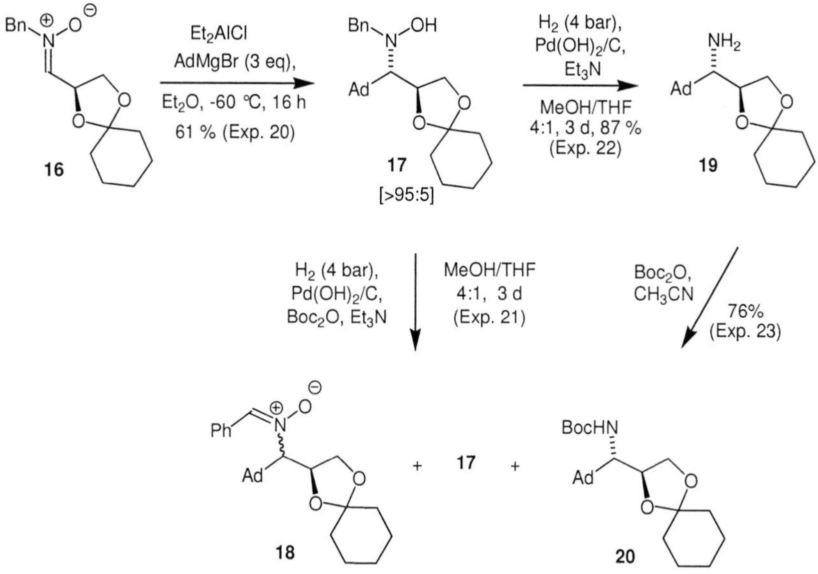

2.5.2 Formation of a nitrone by-product

The addition of adamantylmagnesium bromide to the 2,3-*O*-isopropylidene-D-glyceraldehyde *N*-benzylnitrone **4** and the subsequent hydrogenation have been repeated. This led to an unexpected result. After hydrogenation for 3 days under the usual conditions, 63 % of starting material (with small impurities) was recovered as well as a new compound, (2S,3S)-3-(1-adamantyl)-3-[*N*-(Z)-benzylidene-*N*-oxyamino]-1,2-isopropylidene-1,2-propanediol **(14)**, in 23 % yield (Scheme 8). The NMR data did not correspond to the expected *N*-Boc protected amine **13**, and the structure could only be determined with certainty by means of X-ray crystal structure analysis (Figure 2).

The mechanism of the formation of this nitrone, which involves oxidation, is not clear, and similar examples were not found in the literature.

Figure 2. X-ray diffraction analysis of crystal structure of the nitrone **14**

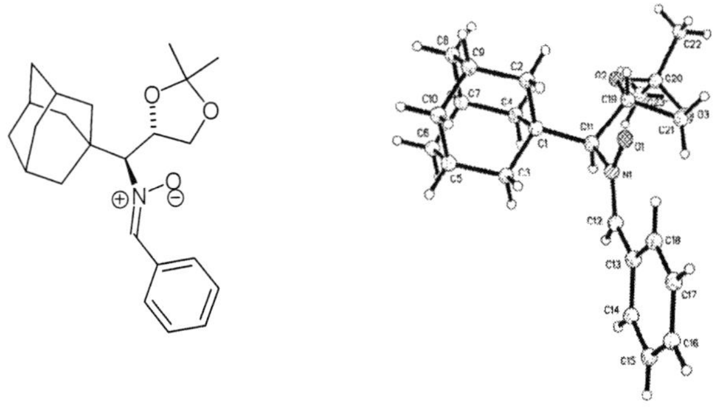

The formation of this byproduct was also observed in the case of the *N*-benzylhydroxylamine **17** having the cyclohexylidene protecting group. The hydrogenation of **17** gave a mixture of the required *N*-Boc protected amine **20**, of the starting material **17** as well as of **18**. The presence of the latter compound was verified by comparison of its NMR and IR data with those of the nitrone **14**. The benzylidene proton of both nitrones **14** and **18** has a characteristic chemical shift (singlet) at 7.32 ppm (in CDCl$_3$). It has been shown that the oxidation of the pure hydroxylamine **17** takes place even in a closed NMR tube by standing in CDCl$_3$ (conversion ca. 30 % after 2 days, ca. 40 % after 5 days). On the other hand, a qualitative test showed that stirring of *N*-benzylhydroxylamine **17** in ethyl acetate* exposed to air did not lead to the formation of the nitrone after 24 h, but after the addition of silica gel or triethylamine traces of the byproduct **18** could be detected (NMR control). These observations let one guess that the oxidation of the main product (and eventually decrease of the yield) happens during chromatographic purification of the product **17** on the silica gel. This conclusion is not in accordance with the observation that standing of the sample in an NMR tube in CD$_2$Cl$_2$ (which is not as acidic compared to CDCl$_3$) showed a conversion of ca. 30 % after 5 days. So far, the conditions of this side-reaction are not clear.

---

* Ethyl acetate was chosen as it is the common solvent for the purification of the compound discussed (in a mixture with petroleum ether).

In practice this would mean that the purification of the addition product **17** on silica gel should be done as fast as possible (optionally without addition of triethylamine).

2.5.3 Removal of the cyclohexylidene protecting group

It has been shown above during the preparation of phenylglycine, that the isopropylidene protecting group could be removed in the presence of a catalytic amount of *p*-toluenesulfonic acid in methanol, but the *N*-Boc protecting group was not affected under these conditions.[47] On the other hand, removal of the cyclohexylidene protecting group needs conditions, which are critical for the *N*-Boc protection, such as aqueous trifluoroacetic[67] or hydrochloric[68] acid. For this reason, the product of the addition **17** in analogy with **7b** was exposed to hydrogenation in the presence of Pearlman's catalyst,[52,53] but without addition of the di-*tert*-butyl dicarbonate. The corresponding amine **19** was obtained in 87 % yield (Scheme 9) and then treated with 12 N HCl to release the 3-amino-1,2-diol hydrochloride **21** (82 % yield). The structure of this compond was confirmed by X-ray crystal structure analysis.* The analysis showed that a molecule of the hydrochloride **21** coordinates to one molecule of water. This fact was however not confirmed by the elemental analysis. The explanation for this might be the additional drying of the hydrochloric salt **21** sample directly before the elemental analysis performance.

This hydrochloride **21** was then treated with di-*tert*-butyl dicarbonate and gave the corresponding *N*-protected aminodiol **22** in 79 % yield (Scheme 10).

Scheme 10. Synthesis of (S)-N-Boc-1-adamantylglycine

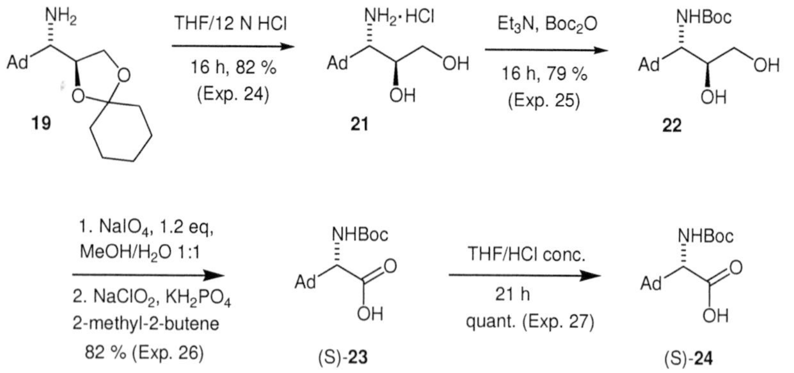

The transformation of the diol moiety[61,62] was done by oxidative cleavage with sodium periodate in methanol-water mixture followed by oxidation to the carboxylic acid with sodium chlorite, as described above for phenylglycine. This procedure afforded N-Boc-1-adamantylglycine [(S)-**23**] in good yield (82 %). The absolute configuration of this amino acid was postulated on the basis of the X-ray crystal structure analysis of the salt **21** as well as by comparison of the ^{1}H NMR data with literature data.[32] The optical rotation value unfortunately was not given in the literature.[32] The configuration of N-Boc 1-adamantylglycine allowed to conclude that the main product of the addition of adamantylmagnesium bromide to the nitrone **16** was the *erythro*(*anti*)-product **17**. This observation was in accordance with the results of the addition of PhMgBr to the nitrone **4** in the presence of Et_2AlCl (see Table 6).

The last step was to remove the N-Boc protecting group with formation of the hydrochloride salt of 1-adamantylglycine: for this, the protected acid **23** was dissolved in THF and treated with concentrated hydrochloric acid at ambient temperature. The crude hydrochloride salt gave the required 1-adamantylglycine hydrochloride (**24**) after recrystallization from methanol/ether mixture. The absolute configuration (S) was confirmed by X-ray analysis. This analysis[†] (Figure 3) also showed that two molecules of glycine hydrochloride share one molecule of methanol from the solvent. This was further confirmed by the elemental analysis. The calculated and found values for the product (S)-**24** (Experiment 8, AB 270) were the following:

	AdGly·HCl $C_{12}H_{20}ClNO_2$	AdGly·HCl·MeOH $C_{13}H_{24}ClNO_3$	AdGly·HCl·1/2MeOH $C_{12.5}H_{22}ClNO_{2.5}$	Found values
C	58.65	56.21	57.35	57.91
H	8.20	8.71	8.47	8.22
N	5.70	5.04	5.35	5.37
Cl	14.43	12.76	13.54	14.34

[†] Anomalous dispersion method was used.

Figure 3. X-ray diffraction analysis of crystal structure of (S)-1-adamantylglycine [(S)-**24**·HCl]

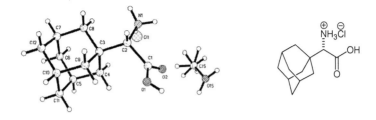

2.5.4   Discussion of NMR spectra of adamantyl and/or cyclohexylidene containing compounds

Looking at NMR data of the products of the addition of phenyllithium, phenyl- and adamantylmagnesium bromides to the nitrones **4** and **16**, one concludes that the assignment of the signals of the atoms in the main backbone (C^1H_2–C^2H–C^3H) is trivial and can be done by means of DEPT and C,H COSY experiments.

The assignment of signals of adamantyl and cyclohexylidene moieties in ^1H as well as ^{13}C spectra is not that simple, it needs a systematic analysis of all compounds from the adamantane series jointly and individually. This approach is similar to a puzzle requiring some assumptions and even speculations at beginning, and at the end leading to the complete pattern of NMR signals.

For all ^1H NMR spectra of compounds containing an adamantyl residue (**7b**, **13-15**, **17-24**), a broad singlet at around 2 ppm was always found (Figure 4). It could be referred to the protons on the C-3', C-5' and C-7'. The correlation of this signal with a signal of a tertiary carbon atom in the ^{13}C NMR spectrum (C,H COSY) was in accordance with this assumption. For all compounds, this signal appears in the range of 28.3-29.5 ppm and sometimes overlaps with those of the three methyl groups of the *tert*-butyl group (28.2-28.4 ppm).

The next step was to distinguish the signals of the methylene groups from the adamantyl and cyclohexylidene residues. First, the comparison of ^{13}C NMR spectra of compounds containing cyclohexylidene residue (**17, 19** and **20**) and and without this diol protection (**21, 22**) was necessary to determine the position of the signals belonging to adamantyl and cyclohexylidene. Then, an estimation of the signal ratios allowed differing them from each other.

Figure 4. A typical fragment of an ¹H NMR spectra of N-Boc aminodiol **20**

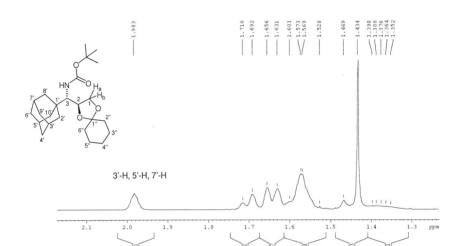

The adamantyl rest is a highly symmetric structure, which led to the assumption that the groups of C-2', C-8' and C-10' as well as C-4', C-6' and C-9', respectively, would give a single signal in the ¹³C NMR spectrum. On the other hand, it is evident and known from the literature that the carbon atoms of a monosubstituted cyclohexane are not equivalent.[69,70]

All this allowed to conclude that the signals of higher intensity around 37 ppm and 40 ppm belong to the adamantane skeleton. Thus, the residual five signals could be assigned as belonging to the cyclohexylidene part (Figs. 5 and 6).

The signals of C-2" and C-6" are very close but not equivalent, and can be found in the 34.7–36.6 ppm region of the ¹³C NMR spectrum. It was not possible to assign these unambiguously. The signals of C-3", C-4" and C-5" as expected were found in the high-field part of the spectra. The signals of C-3" and C-5" appeared closely (or overlapping), whereas the separately appearing signal was identified as that from C-4".

Two signals of carbon atoms of adamantane part could be differentiated bearing in mind that the atoms C-2', C-8' and C-10' should "feel" the changes (the presence of the i.e. substituent at C-1') more than those of C-4', C-6' and C-9'. In fact, comparison of the ¹³C NMR spectra of the compounds **7b** and **13**, **17** and **20** proves that the Δδ values decreased while moving from signals of C-1' to those of C-4', C-6' and C-9'. This observation let to deduce that the signal at high field (38.5–40.4 ppm) belongs to the C-2', C-8' and C-10' atoms, and the low field signal (36.6–37.2 ppm) refers to the C-4', C-6' and C-9' group.

The ^{13}C NMR spectra of compounds **7b**, **13**, **17**, and **19-21** are presented in the Appendix.

Figure 5. ^{13}C NMR spectra of **20**, recorded in $CDCl_3$

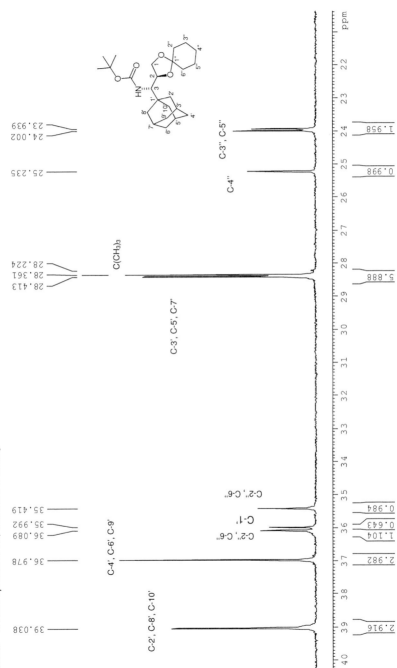

Figure 6. ^{13}C NMR spectra of **22**, recorded in DMSO-d_6

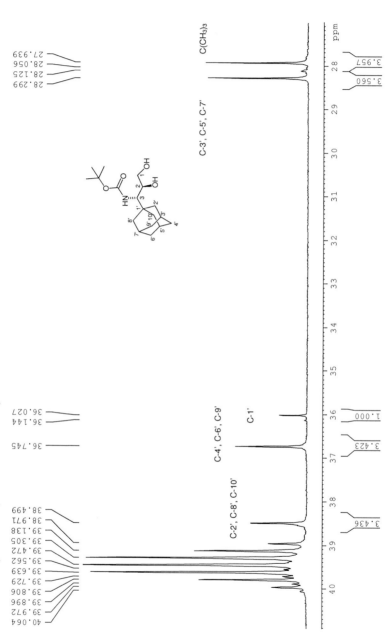

Table 9. ^{13}C NMR data of some adamantyl containing compounds (only aliphatic part is presented)[a]

Compound	C-1'	C-2', C-8', C-10'	C-3', C-5', C-7'	C-4', C-6', C-9'	C-2"/C-6"	C-3"/C-5"	C-4"	C(CH$_3$)$_3$
7b	38.0	40.4	28.7	37.1	-	-	-	-
13	35.9 Δδ=-2.1	39.1 Δδ=-1.3	28.4 Δδ=-0.3	36.9 Δδ=-0.2	-	-	-	28.3
17	38.0	40.4	28.8	37.2	35.1/36.6	24.1	25.3	-
20	36.0 Δδ=-2.0	39.0 Δδ=-1.4	28.3 Δδ=-0.1	37.0 Δδ=-0.2	35.4/36.1	23.9/24.0	25.2	28.3
19	35.1	39.2	28.4	37.1	34.7/36.2	23.8/24.0	25.2	-
21	35.5	39.5	29.5	37.5	-	-	-	-
22	35.5	39.6	28.4	36.8	-	-	-	28.4
23	36.1	38.5	28.3	36.6	-	-	-	28.2

[a]All spectra have been recorded in CDCl$_3$ as a solvent, at 75 or 125 MHz.

2.6   Conclusion

A new stereoselective approach to 1-adamantylglycine has been developed. (S)-1-adamantylglycine hydrochloride [(S)-**24**] was prepared in 6 steps with a yield of 28 % starting from the nitrone **16** (Scheme 11). This approach includes a highly stereoselective addition step, and the selectivity can be controlled by the choice of the Lewis acid applied. This means, both enantiomers of substituted glycines are potentially accessible by this route.

As preliminary experiments showed, this new route can easily be used for the preparation of other substituted glycines with bulky substituents, such as phenyl, tert-butyl and similar.

Scheme 11. Presentation of the total synthesis of (S)-1-adamantylglycine hydrochloride

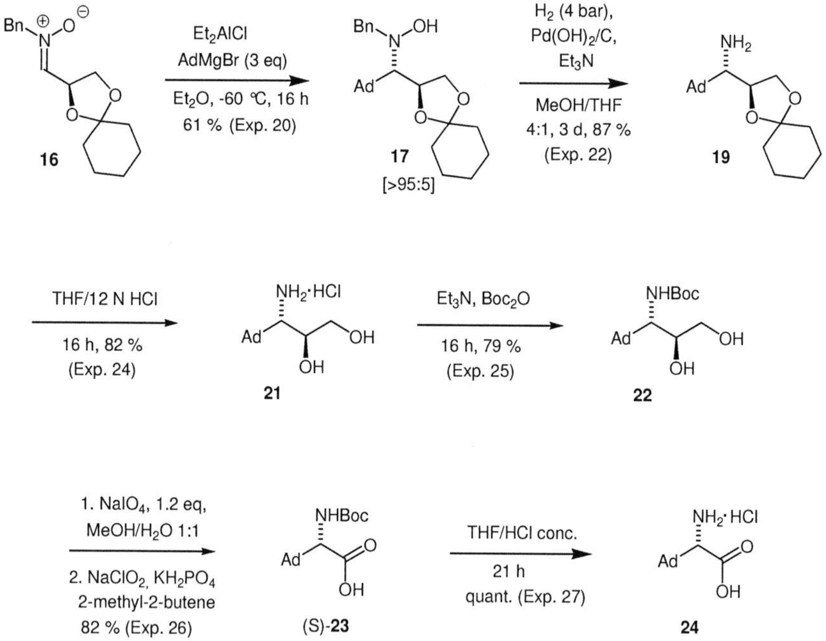

3   SYNTHESIS OF β-AMINO ACIDS WITH BULKY SUBSTITUENTS IN THE β-POSITION

3.1   Background

A countless number of publications and reviews devoted to elaborate methods for the synthesis of various $\beta^{3,3}$-amino acids can be found.[10,15,11,71,72,73,74,75]

According to different classification methods,[10,11] several main approaches for stereoselective synthesis of β-amino acids are available: homologation of α-amino acids (Arndt-Eistert homologation),[76] enzymatic resolution, addition of enolates to imines, Curtius rearrangement, conjugate addition of a N-nucleophile to α,β-unsaturated esters and nitriles, addition of C-nucleophiles to chiral imines, hydrogenation, amino hydroxylation and β-lactam synthesis.[10a,77]

Seebach and Abele[7] suggested an alternative presentation of pathways for the synthesis of achiral β-amino-β,β-disubstituted amino acids (Scheme 12).

Scheme 12. Various methods for the preparation of achiral $\beta^{3,3}$-amino acids[7]

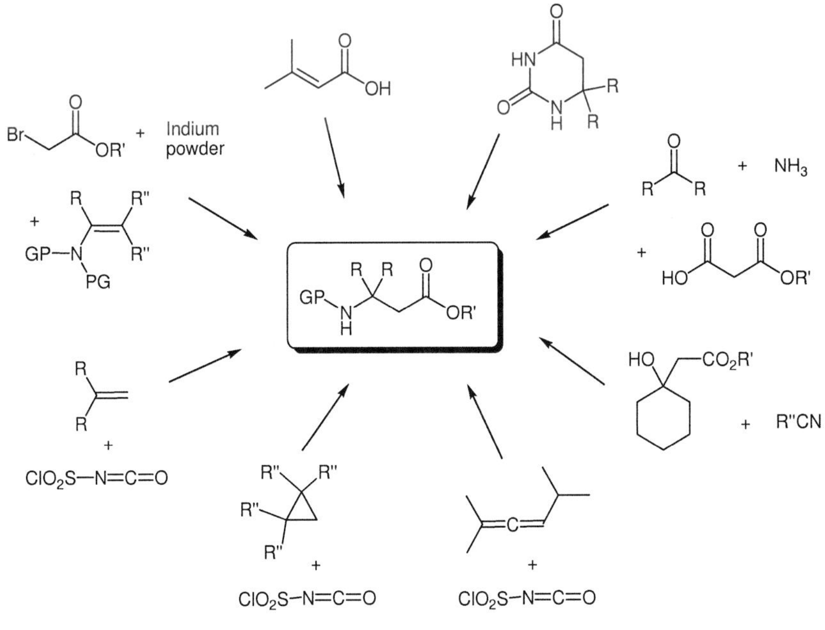

The classifications of the synthetic methods for the preparation of chiral and achiral amino acids presented do not pretend to be comprehensive and serve just for orientation in the plenty of approaches.

Some examples of stereoselective synthesis of sterically constrained β-amino acids are briefly commented below.

3.1.1 Isoxazoline approach to β-amino acids

The synthesis is initiated by a 1,3-dipolar cycloaddition of various substituted oximes **AC** and allylic alcohols **AD**,[‡78] containing the requisite stereochemical information and substitution pattern for ready transformation to β-amino acids. The nucleophilic addition to the C=N bond of the isoxazoline followed by cleavage of the N–O bond and further oxidative cleavage of the resulting diol would then provide the target β-amino acid (Scheme 13)

Scheme 13. Isoxazoline approach to ß-amino acids[15]

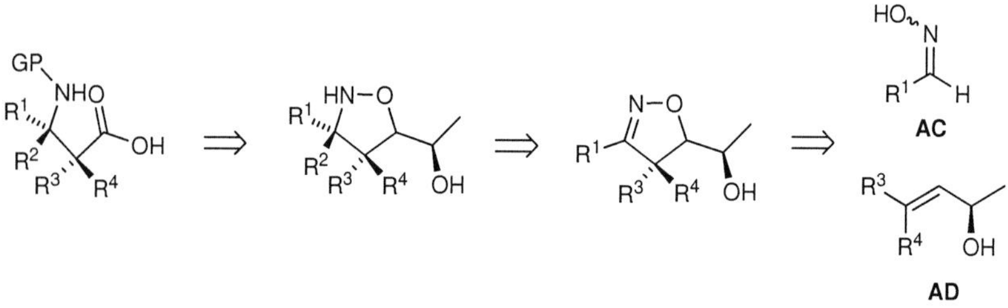

Highly substituted β-amino acids have been prepared according to this route. The results are presented in the Table 10.

Table 10. Synthesis of highly substituted β-amino acids[15]

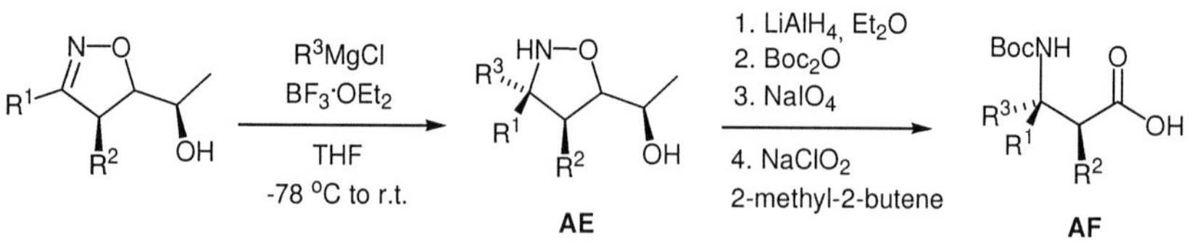

[‡] Both enantiomers of the allylic alcohols of type **AD** (R^3, R^4 = Me and/or H) were prepared starting from acetaldehyde in 27-30 % yield over 3 steps.

R^1	R^2	R^3	Yield of AE [%]	Yield of AF [%]	d.r.
i-Bu	H	allyl	90	54	90 : 10
Et	H	allyl	95	69	91 : 9
i-Bu	H	benzyl	90	54	95 : 5.
Et	Ph	allyl	81	48	95 : 5

3.1.2 Asymmetric addition of O-silyl ester enolates to imines

The asymmetric Mannich-type reaction provides access to β-amino esters. It involves the condensation of an imine and an ester enolate. However, such type of reactions is less applicable due to the poor electrophilicity of the imine function, and its tendency to undergo α-deprotonation leading to the formation of an enamine. One way to avoid these problems is the use of preformed ester enolates, such as the O-silyl derivatives as nucleophiles.[11,71,72,75,79]

Jacobsen et al. described a route to N-Boc-protected β-amino acids through the enantioselective addition of O-silyl ketene acetals to N-Boc-aldimines catalysed by the optically active thiourea catalyst **AG**.[71]

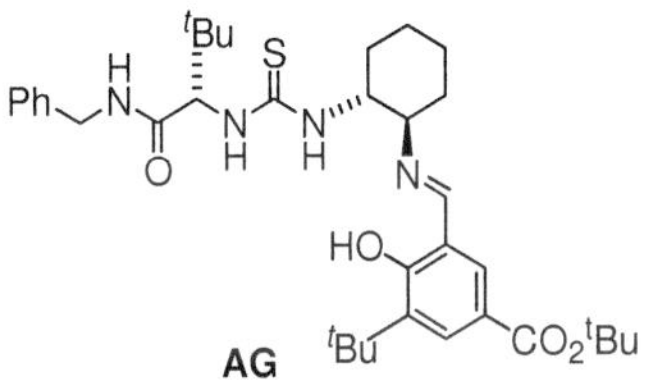

Variation of the silyl and alkoxy groups of the silyl ketene acetal serving as nucleophile led to additional rate enhancement. The use of O-tert-butyldimethylsilyl ketene acetals gave the best results (Table 11). An additional accelerating effect was observed with larger alkoxy substituents (entries 1-3), although the tendency was reversed in the case of the tert-butoxy derivative (entry 5).

Table 11. Optimization of the O-silyl ketene acetal structure in the Mannich type reaction[71]

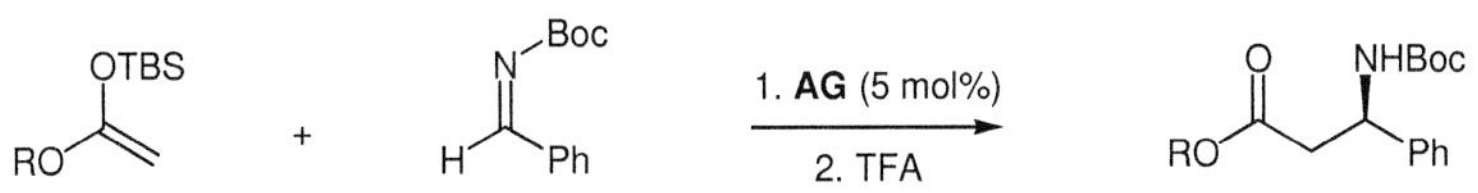

Entry	R	Temp. [$^{\circ}$C]	Time [h]	Conversion [%]	ee [%]
1	Me	23	5.5	90	54
2	Et	23	3.5	90	63
3	i-Pr	23	2.0	93	68
4	i-Pr	−40	48	90	91
5	t-Bu	23	21.5	91	51

A similar approach in the synthesis of β-amino esters was used in the group of Skrydstrup.[72] The Lewis acid mediated addition of various O-silyl ester enolates to N-acylhydrazones was investigated (Scheme 14).

Scheme 14. Reaction of O-silyl ketene acetals with N-acylhydrazones[72]

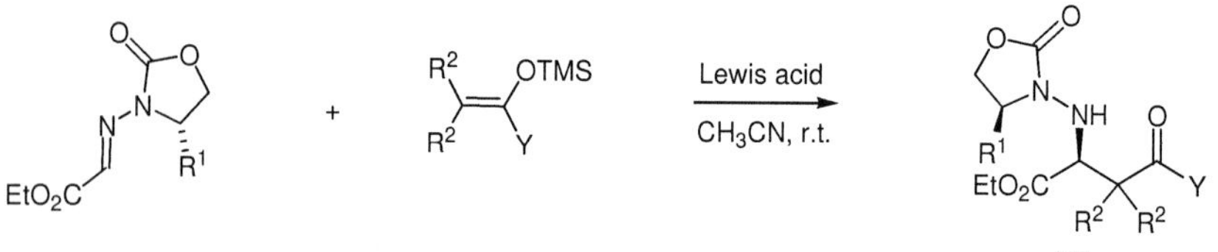

The authors could show the influence of the Lewis acid, of the structure of silyl enolates and of the type of the chiral 2-oxazolidinones employed on the yields of the product **AH** and on the diastereoselectivity of the addition.

3.1.3 β-Lactam approach

It is well known that β-amino acids are precursors of β-lactams. At the same time, this type of amino acids can be prepared from β-lactams. These two classes of compounds are "chemical relatives" since they can be transformed into each other. The interest to the β-lactam approach as a possibility to prepare the corresponding β-amino acids has been driven by the accessibility of enantiomerically pure β-lactams.

Palomo *et al.*[80] presented the 2-azetidinone framework as an effective tool for the construction of a wide variety of structurally different products. The selective ring-opening of β-lactams gives not only β-amino acids, but may also lead to α-amino acids and even peptides.

β-Lactams act as formal acylating agents towards those nucleophiles that effect the cleavage of the N1–C2 bond of an 2-azetidinone (Scheme 15). The majority of the standard procedures for such cleavages deals with hydrolysis (acidic[17] or basic[81,82]) or alcoholysis.[83] In the case of β-lactam systems bearing acid- and/or base-sensitive substituents, alternative methods have to be applied. These could be enzymatic methods as well as application of O-, N-, and C-nucleophiles. For the latter three cases, the importance of N-acyl or N-carbamate (i.e. N-Boc) protected β-lactams is stated. The N-Boc group serves to activate the β-lactam carbonyl towards nucleophilic attack. On the other hand, it protects the amino function of the resulting β-amino acid derivative.[80]

Scheme 15 N1–C2 bond cleavage leading to β-amino acids[80]

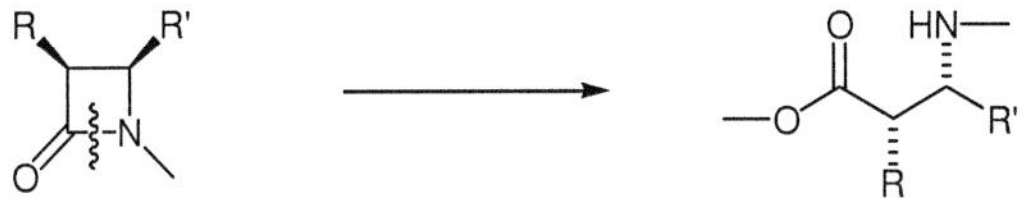

3.1.4 Arndt-Eistert homologation

Arndt-Eistert homologation is a very interesting approach due to the variety of methods for the synthesis of the α-amino acids. It consists in the introduction of a methylene fragment into a protected α-amino acid, whereas no racemisation is usually observed and the corresponding β-amino acid retains the configuration:[84]

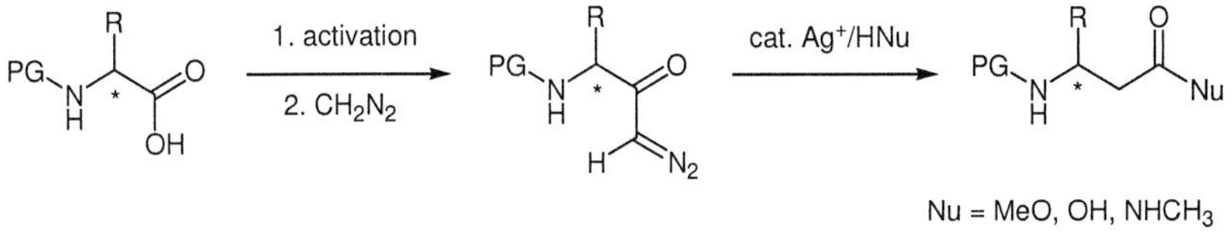

Nu = MeO, OH, NHCH$_3$

Carbamate protecting groups (Z, Boc) are suitable in this procedure. The starting carboxylic acid is converted into the mixed anhydride with, for example, the triethylamine/ethyl chloroformate system, and then treated with diazomethane to yield an α-amino diazoketone. This derivative can be purified and subjected to the subsequent Wolff rearragement in the presence of nucleophiles such as methanol, water, or methylamine, with silver salts as catalyst.

3.2   β-Amino acids *via* homoallylic amines

3.2.1   Background

The approach to β-amino acids presented in the following section can be classified as addition of *C*-nucleophiles to imines. It consists in the addition of allyl reagents to imines followed by oxidative transformations of the allyl group resulting in β-amino acids with the required substituents R^1 and R^2:

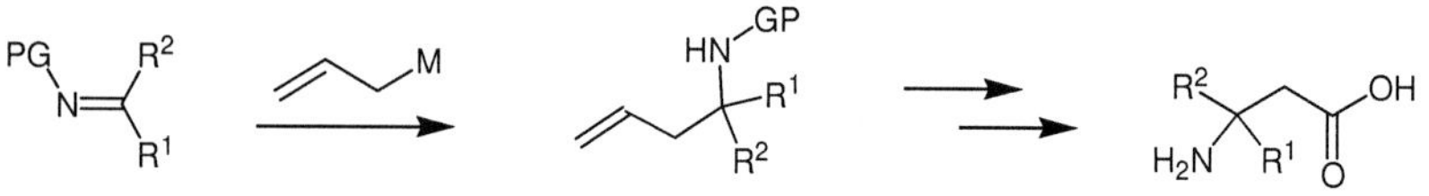

R^1, R^2 = alkyl/aryl rests
M = MgBr, MgCl or Li

Organometallic reagents add to the imine C=N bond giving amine derivatives as a racemic mixture (in case when $R^1 \neq R^2$). The stereoselectivity of the addition can be controlled, as it has already been discussed in the Chapter 2.3. At first glance, this route looks promising for the preparation of amino acids **E-I**. The starting materials for this path could be the following imines, which are accessible from the corresponding ketones:

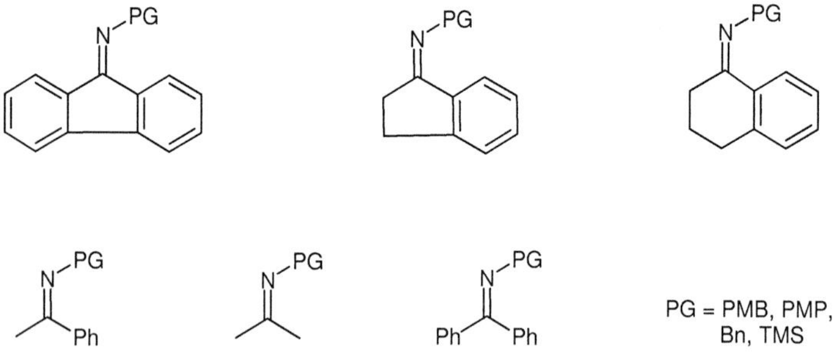

PG = PMB, PMP,
Bn, TMS

The *N*-trimethylsilyl protection group[85,86,87] can especially be interesting in some cases: the N–Si bond is quite labile, therefore the free amine can be obtained after addition of the organometallic compound just by hydrolysis.[88]

An essential requirement of the addition of organometallic compounds (in our case – allylmetallic reagents) is the high stereoselectivity. One of the techniques known is the application of allylboron reagents.[89,90,91,92,93] In 1986 Yamamoto *et al.*[90] reported on highly stereoselective additions to 9-allyl-9-borabicyclo[3.3.1]nonane (9-BBN) to the chiral aldimines **AI**. The results are presented in Table 12.

*Table 12. Selectivity of the allylation of the imine **AI**[90]*

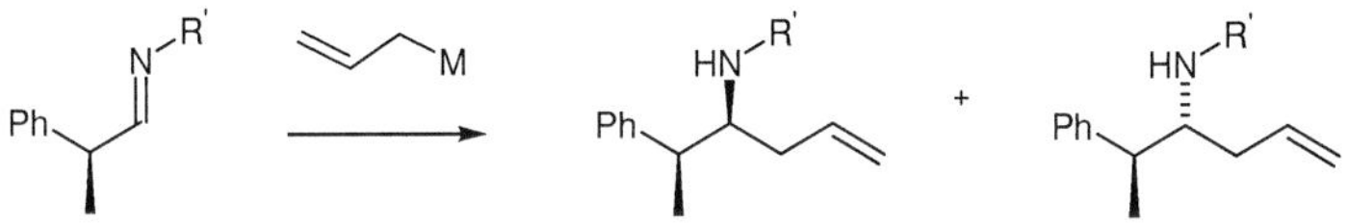

Entry	R'	M	AJa : AJb
1	*n*-Pr	9-BBN	*96 : 4*
2	*i*-Pr	9-BBN	*"100 : 0"*
3	*n*-Pr	SnBu$_3$/TiCl$_4$	*93 : 7*
4	*i*-Pr	SnBu$_3$/TiCl$_4$	*92 : 8*
5	*n*-Pr	MgCl	*84 : 16*
6	*n*-Pr	MgBr	*68 : 32*
7	*i*-Pr	MgCl	*70 : 30*

Although the reaction of imines with allylmagnesium compounds exhibited low selectivities (entries 5-7), the reaction of allyl-9-BBN gave the allyl derivative **AJa** either exclusively or predominantly. The high stereoselectivity was rationalized by the authors by means of the six-membered chairlike transition state, where 9-allyl-9-BBN plays an important role. This transition state led to the preferable formation of the Cram's isomer (*threo*).[90]

Another example of an allylborating reagent is *B*-allyldiisopinocampheylborane [(Ipc)$_2$*B*-allyl], prepared from either *B*-chlorodiisopinocampheylborane (DIP-chloride[TM]) or from *B*-methoxydiisopinocampheylborane and allyl Grignard reagent.[92,94] Originally, this borane was used to provide high diastereoselectivity in the reduction of aldehydes to optically active alcohols.[92] Later it has been shown that this reagent provides high diastereoselectivity in the preparation of homoallylic amines by allylboration of *N*-silylimines (eq. 6).[92]

$$R=Ph, 92\% \text{ ee}$$
$$R=2\text{-ClC}_6H_4, 82\% \text{ ee}$$
$$R=4\text{-MeOC}_6H_4, 92\% \text{ ee}$$

(6)

3.3 Synthesis of β-amino-β,β-diphenylpropionic acid (F)

3.3.1 Use of benzophenone N-(p-anisylmethyl)-imine (25), scope and limitations

Benzophenone has served as a starting material for the synthesis of β,β-diphenyl substituted amino acid **29** as a direct precursor of the target amino acid **F** (Scheme 16). To prepare the protected imine **25**, benzophenone and p-methoxybenzyl amine were heated to reflux in toluene in the presence of $ZnBr_2$ as a Lewis acid with azeotropic destillation of water.[95] The yield of the imine product **25** was 57 % as compared to 71 % reported in the literature.[95] The treatment of the imine **25** with allylmagnesium bromide took place smoothly and gave the allyl derivative **26** in excellent yield.

According to the synthesis plan (see Introduction), the following step should be an oxidative transformation of the allylic moiety which should lead to the N-p-methoxybenzyl-protected amino acid **28**. At this step, some methods for the oxidation of the double bond of **26** to the carboxy group have been tested.

Initially, the reaction of oxidation of the double bond with sodium periodate in the presence of a catalytic amount of ruthenium trichloride[44] as well as oxidation by means of oxone in the presence of the catalyst osmium tetroxide[96] have been carried out. The experiments failed to give the required product. Decomposition of the starting material (TLC control) was observed in both cases. Another typical method – oxidation with potassium permanganate[97] – gave an unknown crystalline product in "24 %" yield along with decomposition products. The structure of this solid was not established.

These results showed the necessity of the transformation of the allyl derivative **26** into the 1,2-diol product **27**. The oxidative cleavage of the dihydroxy moiety proved to be a reliable procedure in the syntheses of substituted glycines (Chapter 2).

For the preparation of the diol **27** two methods have been used (A, B, Scheme 16). The reaction with osmium tetroxide in the presence of *N*-methylmorpholine *N*-oxide as reoxidizing agent[98] afforded the required compound in only 39 % yield, which was not satisfactory. The diol **27** was then obtained in a better yield (74 %) using potassium osmate dihydrate as a catalyst and potassium hexacyanoferrate as reoxidant (Sharpless conditions).[99] The second method was rather slow and needed five days to be complete, as compared with method A.

The transformation of the diol moiety[100] was done by oxidative cleavage with sodium metaperiodate in methanol-water mixture to give the corresponding aldehyde, followed by *in situ* oxidation to the carboxylic acid **28** with sodium chlorite. The structure of the 3-(4-methoxybenzylamino)-3,3-diphenylpropanoic acid (**28**) was not confirmed. The NMR spectra could not be recorded since the product was soluble neither in water nor in an organic solvent. In order to purify this product, an attempt to change the *N*-protecting group was done. For the removal of *p*-methoxybenzyl (as well as *p*-methoxyphenyl) protecting group ammonium ceric(IV) nitrate (CAN) is usually used.[101,102] In order to isolate the free 3-amino-3,3-diphenylpropanoic acid, ion-exchange chromatography (resin, H$^+$-form) was applied.[103] The purification was repeated twice, but did not succeed, the elemental analysis of the title compound was not correct. This acid was not soluble in any solvent to measure NMR-spectra. Treatment with di-*tert*-butyl dicarbonate did not permit to isolate any product.

Since the isolation of *N*-PMB protected amino acid **28** has been problematic because of its very bad solubility, the esterification of the acid **28** *in situ* without isolation was carried out (Scheme 16). Diazomethane as a solution in diethyl ether was used therefore. The methyl ester **29** was isolated without any problem, however in quite a low yield (28 % over 3 steps starting from the diol **27**). The low yield was perhaps the result of the quality of the diazomethane solution and a higher yield might be possible with a carefully prepared solution.

The compound **29** is an in fact the required amino acid **F** with both amino and carboxy functions protected.

Scheme 16. Synthesis of the methyl ester of N-protected β-amino-β,β-diphenylpropionic acid
(29)

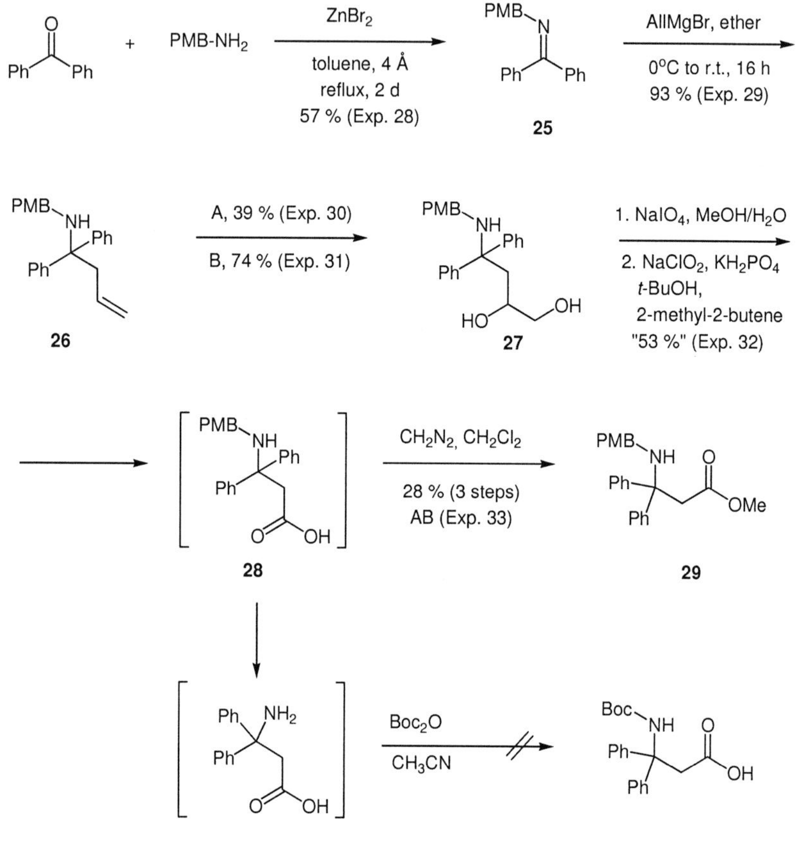

Method A: 1. 10 wt % NMO/H$_2$O, 2.5 wt % OsO$_4$/t-BuOH, t-BuOH/THF, 6 h. - 2. Na$_2$SO$_3$
3. Column chromatography.
Method B: 1. K$_3$Fe(CN)$_6$, K$_2$OsO$_2$(OH)$_4$, K$_2$CO$_3$, t-BuOH/H$_2$O, 5 d
2. Na$_2$SO$_3$ 3. Recrystallization

3.3.2 *N*-Diphenylphosphinylimide as an alternative activating/protecting group

N-Diphenylphosphinoylamides are known to be suitable substrates for nucleophilic addition
to the C=N imine bond. Thus, Schnabel[50] in 1990 reported on the addition of primary

nitroalkanes to the *N*-diphenylphosphinoyl benzaldimine **AK** (Scheme 17). The addition resulted in good yields (70–99 %) and excellent stereoselectivities (*syn:anti* 92:8 to >95:5).

Scheme 17. Reaction of primary nitroalkanes with N-bezylidene diphenylphosphinoylamide[50]

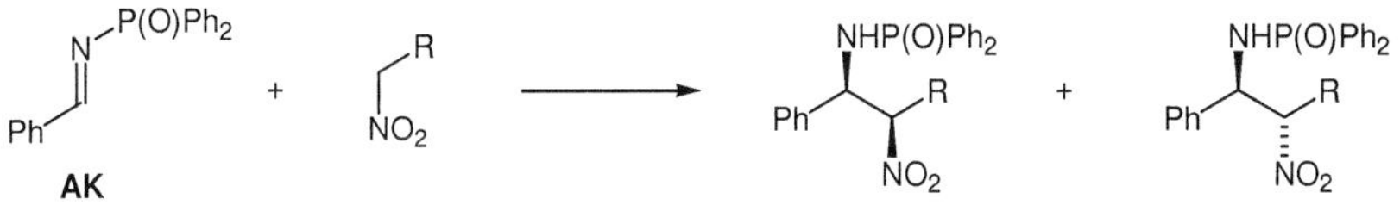

In this work, the *N*-diphenylphosphinoyl compound **31** was applied as imine component. In order to obtain it, benzophenone was treated with hydroxylamine hydrochloride and pyridine in dry ethanol followed by the addition of aqueous ammonia.[104] Benzophenone oxime (**30**) was isolated in 74 % yield and after reaction with diphenylphosphinous chloride gave the required *N*-(diphenylmethylene)diphenylphosphic amide (**31**) in a moderate 42 % yield. The starting oxime **30** was reisolated in 11 % yield inspite of the fact that 1.5 equivalents of diphenylphosphinous chloride were used and the reaction was run for 27 h. Along with **30** and **31**, a by-product, *O*-(diphenylphosphinoyl) benzophenone oxime (**32**), could be isolated (2 %). These results show that this method is not the optimal one to produce the amide **31**. Nevertheless, the allylation of this protected benzimine was untroubled and gave the allyl derivative **33** in 82 % yield. The following step – removal of the *N*-protecting group – was unsuccessful. An attempt to treat the imine **31** with hydrochloric acid[50] followed by neutralization gave a mixture of compounds and the NMR data were uninformative.

In summary, the approach described shows that the *N*-diphenylphosphinoyl protecting group is not suitable here due to the low yields and complications with its removal.

Scheme 18. Application of N,N-diphenylphosphinoyl as a protecting group

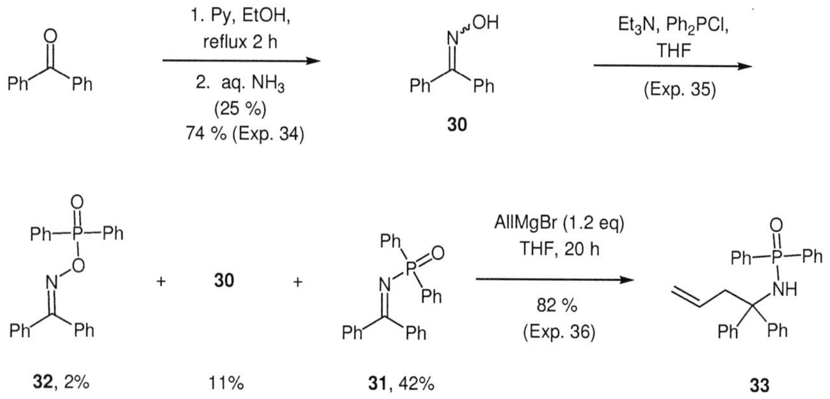

3.3.3 An effort to utilize benzophenone *N*-trimethylsilylimine for the syntheis of β-amino-β,β-diphenylpropionic acid (F)

The difficulties with the isolation of the *N-p*-methoxybenzyl-protected amino acid **28** as well as with the removal of the *N*-diphenylphosphinic protecting group let conclude that a different *N*-protecting group in the initial benzophenone imine help to optimize the sequence leading to the required β-amino-β,β-diphenylpropionic acid (**F**). One of the possibilities is the application of the *N*-trimethylsilyl group. Since the metal-nitrogen bond is easily hydrolyzed, the *N*-metalloimines may be considered a protected, stabilized form of the corresponding imine of ammonia, which have a tendency to trimerize to form triazines.[88] The preparation of such an imine from benzophenone (**AL**, Scheme 19) is described in the literature.[113] According to this procedure, benzophenone in absolute benzene was treated with LiHMDS and heated at 70 °C for 7 h followed by addition of trimethylsilyl chloride. The product of this reaction was not isolated but treated with an ethereal solution of allymagnesium bromide. The result of this sequence should be the free amine **AM**. Instead, the product of the addition of allylmagnesium bromide to benzophenone was isolated, the alcohol **34** (Scheme 20). This result showed that the formation of the imine **AL** had failed. A solution of this problem could be the application of the Ritter reaction[105,106] – transformation of the hydroxy group into the corresponding acetamide moiety, see below.

Scheme 19. Attempted preparation of 1,1-diphenylbut-3-enylamine

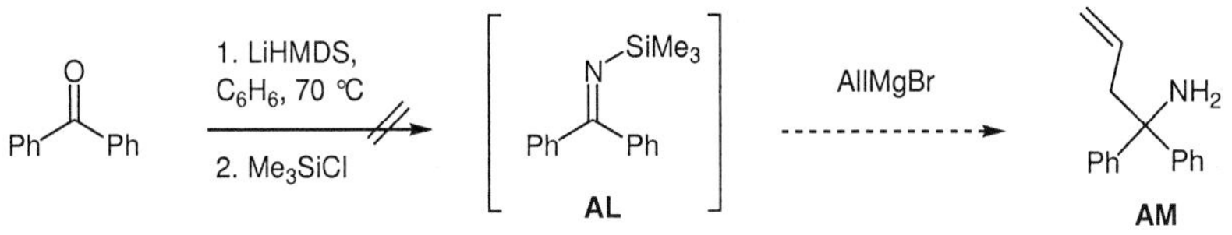

3.3.4 The Ritter reaction as a means to introduce the amino group

The alcohol **34** was synthesized from benzophenone and allylmagnesium bromide in order to subject it to the Ritter reaction (Scheme 20). The resulting allyl derivative **34** was treated with catalytic sulfuric acid in acetonitrile.[105,106] Acetonitrile served both as solvent and source of nitrogen. The product expected was the *N*-acetamide **AN**. However, the compound that was isolated in 41 % yield from the mixture of few compounds was the elimination product, the diene **35**, as followed from the comparison of the NMR data with the literature data.[107] The formation of this product can be explained by generation of the stable 1,1-diphenylbutadiene (**35**). Sulfuric acid might have been too strong for this case. Chlorosulfonic acid was used instead, but also without success; the NMR data of the crude product showed the predominant presence of the diene product **35**.

Some further attempts were done in order to use the alcohol **34** in a modified Ritter reacton[108] (Scheme 20) in which this compound was treated with sodium cyanate in a mixture of sulfuric and acetic acid. According to the literature,[108] an *N*-formyl amide should be the product of this type of transformation. Nevertheless, instead of the expected product **AO** the already known elimination product **35** was obtained once again.

Scheme 20. An attempt to introduce an acetyl group by means of the Ritter reaction

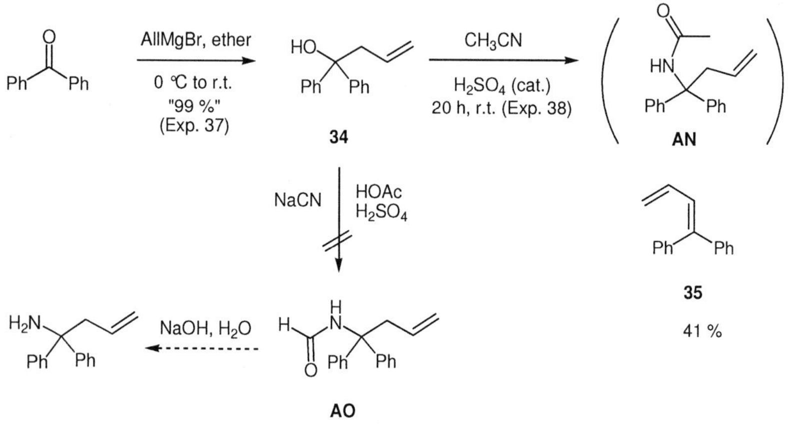

3.3.5 Conclusion

Thus, the target β-amino-β,β-diphenylpropionic acid (**F**) was synthesized in the form of its
N-PMB-protected methyl ester. The sequence presented in Scheme 16 is so far the best
route to this compound, however some optimisation is needed. Alternative methods such as
the application of *N*-diphenylphosphinic or *N*-trimethylsilylimines failed or did not give
satisfactory results. Another possibility would be the way *via* β-lactams, which is described in
the following (Chapter 4).

3.4 Synthesis of 9-aminofluorenylacetic acid (E)

3.4.1 Attempt to use fluorenone *N*-benzylimine for the preparation of
9-aminofluorenylacetic acid (E)

The amino acid **E** was planned to be prepared according to the already known sequence:
introduction of the allyl moiety by means of nucleophilic addition to the imine C=N bond, with
ensuing oxidative degradation of the double bond of the olefine. For the preparation of the
imine **36**, a procedure described in the literature was applied.[109]. Condensation of fluorenone
with benzylamine in the presence of titanium tetrachloride gave the *N*-benzylimine of
fluorenone in good yield (61 %) (Scheme 21). The reaction of this imine with allylmagnesium
bromide ran smoothly, although the following oxidative transformation of the allyl moiety
proved not to be effective. It was not possible to transform the latter directly to the carboxylic

acid by means of sodium periodate in the presence of ruthenium trichloride;[44] instead, decomposition of the starting material was observed (TLC monitoring). An alternative method according to Sharpless showed a better but still unsatisfactory result.[110] The allyl derivative **37** was treated with catalytic amount of potassium osmate dihydrate as well as potassium hexacyanoferrate as reoxidant. Some starting material (23 %) was reisolated after two days. These results forced us to investigate alternative reaction conditions, presented in the following chapter.

*Scheme 21. The route via fluorenone N-benzylimine (**36**)*

36   **37**

3.4.2 Fluorenone *N*-trimethylsilylimine for the synthesis of 9-(*N*-Boc-amino)-fluorenylacetic acid (42)

N-Metalloimines are widely used in organic synthesis for reactions with various nucleophiles. As mentioned above, the *N*-trimethylsilyl group is a suitable one for the reaction between an imine and allylmagnesium bromide. Commonly, *N*-trimethylsilylimines can be obtained by treatment of aldehydes or non-enolizable ketones with lithium hexamethyldisilazide (LiHMDS).[87,111] This reagent can be prepared from 1,1,1,3,3,3-hexamethyldisilazane and *n*-butyllithium prior to use.[112] As our experiments showed, better results were obtained using commercially available LiHMDS as a 1 M solution in THF (supplied by Fluka).

As illustrated in Scheme 22, fluorenone was treated with LiHMDS. The corresponding *N*-trimethylsilyl **38** was not isolated since it is very sensitive to the presence of water.[88,113] The ethereal solution of allylmagnesium bromide was directly added to the THF solution of the fluorenimine.[111] The expected 9-allyl-9-aminofluorene (**39**) was isolated in only 24 % yield. The low yield was obviously related to the long tedious purification: repeated column chromatography was required to obtain the pure amine **39**. The structure of this allyl derivative was confirmed by NMR analyses as well as by mass spectroscopy.

A much better yield was recorded when the crude amine **39** without purification was treated with one equivalent of Boc-anhydride in acetonitrile.[114] By this method, the yield of **40** was

61 % over 3 steps.

3.4.3 Completion of the synthesis of 9-(*N*-Boc-amino)-fluorenylacetic acid

In order to derivatize the double bond of the allyl moiety in compound **40**, the same conditions as used for the allyl derivative **26** (Scheme 16) were applied: a catalytic amount of potassium osmate dihydrate and potassium hexacyanoferrate as oxidant.[110] The reaction proved to be very slow; six days were needed to reach full conversion and good yield (86 %). The transformation of the 1,2-diol moiety followed according to the known procedure: oxidative cleavage with sodium periodate, then oxidation of the corresponding aldehyde in the presence of sodium chlorite.[61,62] The *N*-Boc-protected β-amino acid **42** thus was synthesized in 41 % yield over 6 steps.

*Scheme 22. Synthesis of 9-(N-Boc-amino)-fluorenylacetic acid (**42**)*

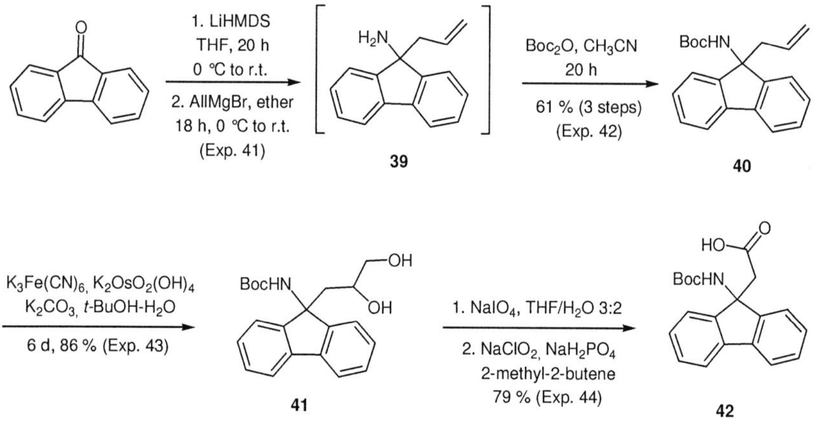

3.4.4 Conclusion

In summary, one of the target amino acids **E** has been prepared as its *N*-Boc derivative, according to Scheme 22. The route includes six steps, but „one-pot" procedures should allow to skip the isolation of some intermediate products and improve the total yield.

3.5   Attempts at the preparation of β-amino-β,β-dimethylpropionic acid (G)

In order to prepare β-amino acid **G**, the method applied for the synthesis of the amino acids **E** and **F** came into consideration, since it had proved to be effective. Acetone served as a starting material in this case, reacting with *p*-methoxybenzylamine in the presence of Na_2SO_4. Acetone was used in a large excess and thus also served as solvent. The imine **43** was isolated in quantitative yield (Scheme 23). The reaction of the imine with ethereal allylmagnesium bromide led to the required allyl derivative, but in moderate yield (47 %). The following attempt to subject the double bond of the allyl moiety to the Sharpless dihydroxylation [catalyst $K_2OsO_2(OH)_4$ and oxidant $K_3Fe(CN)_6$][99] led to decomposition of the starting material (TLC monitoring).

*Scheme 23. Towards the synthesis of β-amino-β,β-dimetylpropionic acid (**G**)*

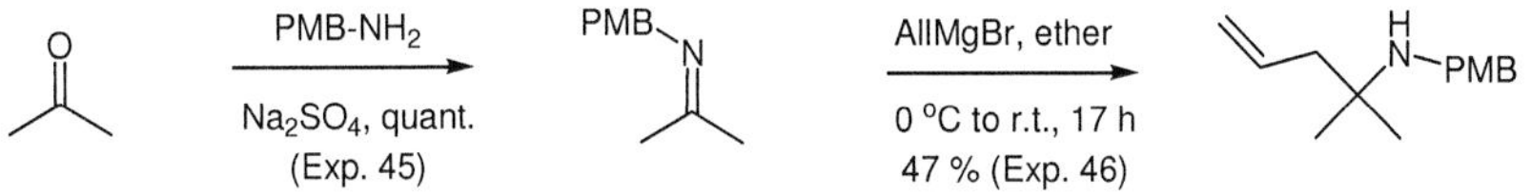

So far, this approach to the β-amino acid **G** was not completed. It is not excluded that the failure of the reactions described above is due to the ability of the imine **43** to undergo deprotonation.[31] In order to prove this point, further experiments with enolizable imines are needed.

Another possibility to reach the required acid **G** and others is the development of alternative methods for the preparation of branched β-amino acids, which will be discussed in the following chapters.

4   β-LACTAMS AS PRECURSORS OF β-AMINO ACIDS

4.1   Background

4.1.1   General remarks

The unsuccessful experiments dealing with the introduction and transformation of the allyl moiety required the development of principally different approaches in the preparation of $\beta^{3,3}$-aminodisubstituted propionic acids. The next logical step could be the combination of a moiety that already contains a carboxy group with an azomethine component. The most suitable method of such a combination might be the addition of O-metal enolates to the C=N bond of an imine.

This type of addition, closely related to the Staudinger reaction, is well-known in the literature and it has been shown that such reactions between ester enolates and imines can lead either to an open-chain β-amino ester or to a β-lactam (Scheme 24).[115]

Scheme 24. The ester enolate – imine condensation[115]

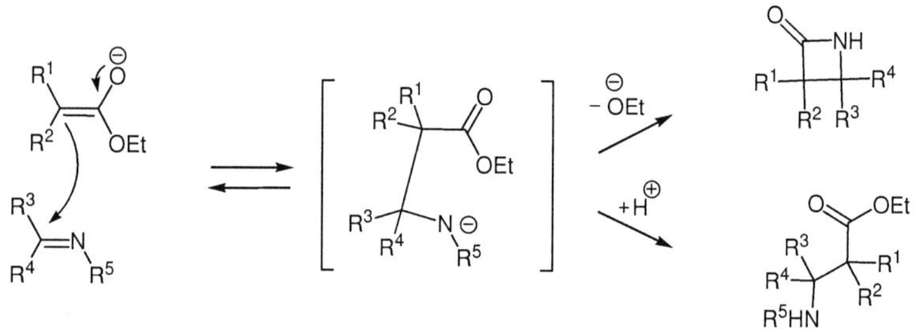

These two products virtually correspond to the same, since the β-lactam is generated from the β-amino ester after cyclization. Exploring this way to β-amino acids, both outcomes of the reaction would be interesting. First, both β-amino acids and β-lactams are classes of chemical compounds possessing interesting biological properties. On the other side, having a β-lactam, the corresponding β-amino acid can readily be obtained (see Chapter 3.1.3).

4.1.2   Literature review

Hart *et al.* introduced a variant of the ester–aldimine condensation that affords β-lactams.[116] Using various esters and imines, appropriate C-3 and C-4 substituents can be introduced into the 2-azetidinone ring. It has been demonstrated that the yield depended on the type of substituent R^1 of the ester enolate (Table 13).

Table 13. β-Lactams from ester enolates and imines[116]

R^1⌒CO_2Et

1. LDA, THF
2. $PhCH=NR^2$
3. HCl, H_2O

→ **AP** + **AQ**

Entry	R^1	R^2	R^3	AP [%]	AQ [%]
1	H	$SiMe_3$	H	14	0
2	Me	$SiMe_3$	H	41	3
3	Et	$SiMe_3$	H	72	0
4	i-Pr	$SiMe_3$	H	80	1
5	t-Bu	$SiMe_3$	H	40	0
6	Me	Ph	Ph	45	2
7	i-Pr	Ph	Ph	87	1

Steric effects proved to be important. Both very small and very large substituents at the α-position of the ester decreased the yield (entries 1, 2, 5) as compared to Et or i-Pr substituents. The principal possibility of synthesis of 3-unsubstituted β-lactams has been demonstrated, whereas it has been stated earlier[117] that ethyl acetate (R^1 = H) and propionate (R^1 = Me) failed to afford β-lactams upon treatment with benzylidene aniline under similar conditions. Concerning the behaviour of N-trimethylsilyl- and N-arylimines, there were no significant differences (entries 2 and 6, 4 and 7).

Fujisawa et al.[118,119] investigated the addition of various ester enolates to an imine **AR** having a chiral 1,3-dioxolane residue (Table 14).Both diastereomers of the β-lactam **AS** could be prepared according to this scheme with excellent diastereoselectivities. The choice of the appropriate metal of the ester enolate allowed to control the selectivity of the addition. Addition of the lithium enolates provided (4S)- β-lactams, while the corresponding (4R)- β-lactams were predominantly obtained by the condensation of titanium enolates with the chiral imine **AR**. The preparation of 3-unsubstituted β-lactams has been demonstrated (R = H), in low yield (42 %) in the case of the lithium enolate of ethyl acetate and in very good yield using the triisopropoxytitanium enolate.[118,119]

Table 14. Addition of various ethyl ester enolates to the chiral imine **AR**[118]

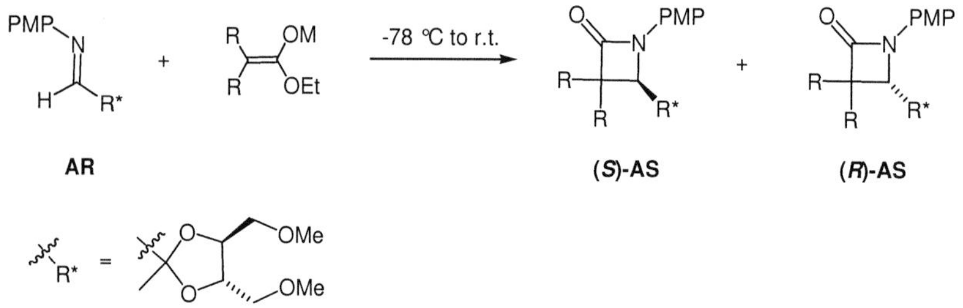

R	M	Solvent	Yield [%]	(S)-**AS** : (R)-**AS**
-(CH$_2$)$_5$-	Li	DME	85[a]	98 : 2
-(CH$_2$)$_5$-	Ti(O^{i}Pr)$_3$	Et$_2$O	82[b]	8 : 92
Et	Li	DME	72[b]	98 : 2
Et	Ti(O^{i}Pr)$_3$	Et$_2$O	76[c]	5 : 95
OEt	Li	DME	55[b]	85 : 15
OEt	Ti(O^{i}Pr)$_3$	Et$_2$O	19[b]	9 : 91
H	Li	DME	42[a]	87 : 13
H	Ti(O^{i}Pr)$_3$	Et$_2$O	85[b]	99 : 1

[a]3 Equiv. of ester enolate were used. - [b]6 Equiv. of ester enolate were used. - [c]15 Equiv. of ester enolate were used.

In the reaction with the enolate of ethyl acetate, the (S)-diastereomer of **AS** formed predominantly regardless of the metal enolate used. In order to investigate this phenomenon, the same research group[119] investigated the addition of the enolates derived from bulky *tert*-butyl acetate. Surprisingly, the β-amino esters **AT** were isolated which did not undergo cyclization to afford β-lactam **AS**. The addition was also very diastereoselective, as shown in Table 15. Addition of lithium or triisopropoxytitanium enolate derived from *t*-butyl acetate to the chiral imine **AR**, possessing dioxolane ring as a chiral auxiliary, gave (3S)-β-amino ester exclusively, whereas chlorozinc enolate gave (3R)-isomer with good selectivity.

The β-amino esters thus obtained were readily converted into the corresponding β-lactams. Hydrolysis of the ester function to the corresponding carboxylic acid was followed by cyclization in the presence of triphenylphosphinedipyridyldisulfide in acetonitrile.[119]

Table 15. Addition of various tert-butyl ester enolates to the chiral imine **AR**[119]

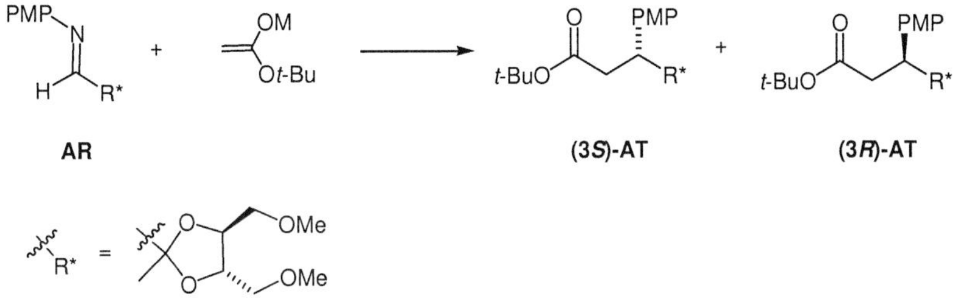

M	Temperature [°C]	Yield [%]	(3S) : (3R)
Li	-78 – -40	68	>99 : 1
Ti(O^{i}Pr)$_3$	-78 – -40	63	>99 : 1
Ti(O^{i}Pr)$_3$	-78 – r. t.	68	>99 : 1
ZnBr	-78 – r. t.	68	14 : 86
ZnCl	-78 – r. t.	77	8 : 92

The preparation and especially the application of various *N*-metalloimines have also been investigated intensively in the group of Cainelli since the 1980's. The authors reported on the application of *N*-trimethysilylimines as key intermediates in the asymmetric synthesis of thienamycin and related carbapenems and penems.[88]

In this work, fluorenone *N*-trimethylsilylimine **38** was chosen as the ketimine component (Scheme 25). The advantages of this imine have been already discussed (Chapter 3.4). The lithium enolate of ethyl acetate (**45**) served as a nucleophile. Since this does not have substituents at the α-carbon atom, the corresponding amino acid will exclusively be substituted at the β-position.

Scheme 25. Planned addition of the lithium enolate of ethyl acetate to fluorenone N-trimethylsilylimine

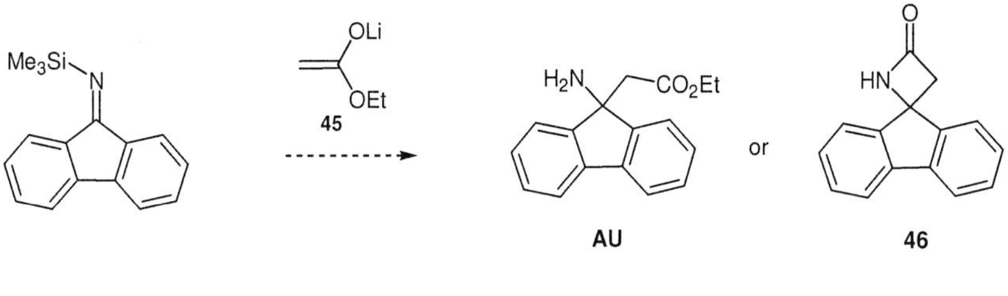

4.2   Experiments on the preparation of spiro[azetidine-2,9'-[9H]fluorene]-4-one (46)

4.2.1   General remarks

In all experiments described in this Chapter, fluorenone was used as the starting material. In order to obtain the corresponding N-trimethylsilylimine, this ketone was treated with a commercially available solution of lithium bis(trimethylsilyl)amide (LiHMDS) in THF according to a procedure described in the literature.[120] The reaction took place in THF at room temperature under nitrogen, for 16–20 h. The corresponding N-trimethylsilylimine was used without isolation in the next experiments shown in Table 16.

The lithium enolate of ethyl acetate (45) was prepared from the corresponding ester by treatment with lithium diisopropylamide (LDA), which was prepared in the previous step from diisopropylamine and n-butyllithium according to the known procedure.[121]

All these reactions did not yield even a trace amount of the β-amino ester **AU**, as deduced from the NMR data of the crude products. For this reason, only the term "azetidine" appears in the name of this chapter.

4.2.2   Experiments towards the β-lactam 46

In the first experiment, the N-trimethylsilylimine **38** was treated with an excess of the lithium enolate of ethyl acetate (1.5 equivalents, entry 1 of Table 16). After hydrolysis of the reaction mixture with a saturated aqueous NH_4Cl solution and appropriate treatment, the β-lactam **46** was isolated albeit in low yield, 11 %. Along with it, 32 % of the starting material (fluorenone) were recovered. The starting ketone was isolated since it is a product of hydrolysis of the N-trimethylsilylimine.[113] The hydrolysis is accelerated under acidic conditions,[122] and the quenching of the reaction with a slightly acidic solution of NH_4Cl could have promoted this process:

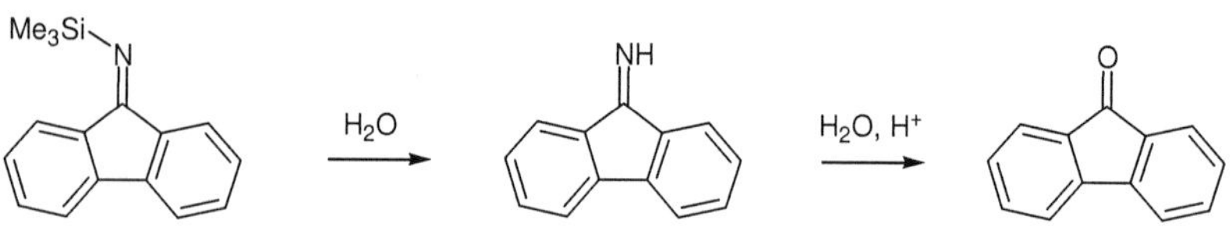

38

The formation of the ß-lactam as a (minor) product of the reaction was not surprising since similar examples are presented in the literature (see discussion in Chapter 4.1.2). The fact that a large part of the starting material was reisolated could indicate that 1.5 equivalents of the lithium enolate of ethyl acetate (**45**) were not sufficient to complete the reaction.

In the following attempt (Table 16, entry 2), ca. 3 equivalents of the lithium enolate **45** were used (here it is the question of using an approximate amount of the lithium enolate, since it is a product of a two-step process). At the same time, a slight excess of LDA was used to assure that all of the ethyl acetate had been consumed. The NMR data of the crude product showed the presence of the required β-lactam product **46** (which was eventually isolated in 40 % yield) in a mixture with some aliphatic compound. This compound was isolated and the NMR data allowed to conclude that ethyl acetoacetate had formed along with the main product of the reaction **46**. The formation of acetoacetate could be the effect of Claisen condensation between the Lithium enolate **45** and the ester (ethyl acetate). Coming back to the NMR data of the crude product, these demonstrated the presence of another product of this reaction – a "dimeric" product **47**. At that moment, this was not isolated nor characterized. This was done afterwards (Table 16, entry 3) and it could *ex post facto* be proven that it presented in the mixture, 25 % of the amount of β-lactamic product **46** (as followed from the ^{1}H NMR spectrum) (entry 2, Table 16).

The next task was to prevent the formation of ethyl acetoacetate. If the idea about the cross-reaction between Lithium enolate of ethyl acetate and ethyl acetate was right, it was necessary to avoid an excess of the ester. As illustrated in Table 16 (entry 3), two equivalents of the ester were used then as well as 2.96 equivalents of the base LDA. As the analysis of the crude product demonstrated, the formation of ethyl acetoacetate had been suppressed. After purification, the 2-azetidinone product **46** was obtained in 34 % yield as well as the dimeric product **47**. The structure of the latter compound was established based on X-ray crystal structural analysis (Fig. 7).

*Figure 7. Crystal structure of the product **47** by X-ray diffraction analysis*

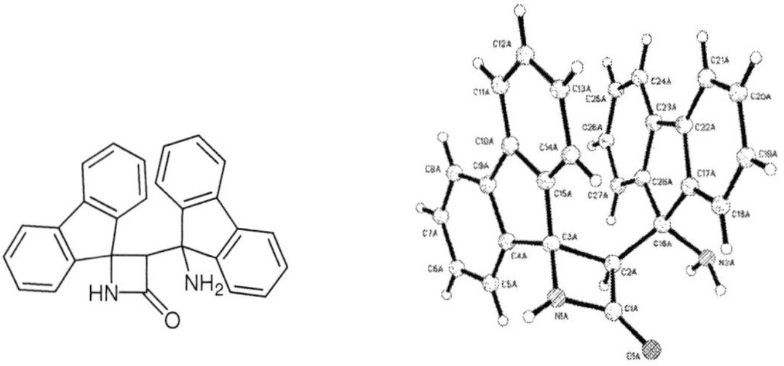

In the experiment discussed (Table 16, entry 3), a large excess of the strong base (LDA) was available. Taking into account that the product **46** is present in the reaction mixture in the form of the *N*-trimethylsilylamide, one can imagine that a proton is abstracted from the acidic methylenic C-3 position of the 2-azetidinone ring. The resulting carbanion could then act as a nucleophile and attack the *N*-trimethylsilylimine **38** during the course of the reaction (Scheme 26).

If this assumption is valid, the absence of a strong base or only a small excess would exclude the possibility of the formation of 3-substituted β-lactam **47**. In order to prove this, the experiment was repeated with 3 equiv. of the lithium enolate **45** and a slight excess of LDA (entry 4, Table 16). The result was exclusively the formation of the "dimeric" product **47** (53 %).

Finally, the reaction was carried out with 1.5 equiv. of the enolate **45**, thus in large excess compared to the strong base (Table 16, entry 5). Under these conditions, none of the required product **46** was detected in the crude product. The starting material was recovered in the form of both fluorenone and fluorenone imine in 39 and 40 % yield, respectively.

*Scheme 26. Proposed mechanism of the formation of the 3-substituted β-lactam **47***

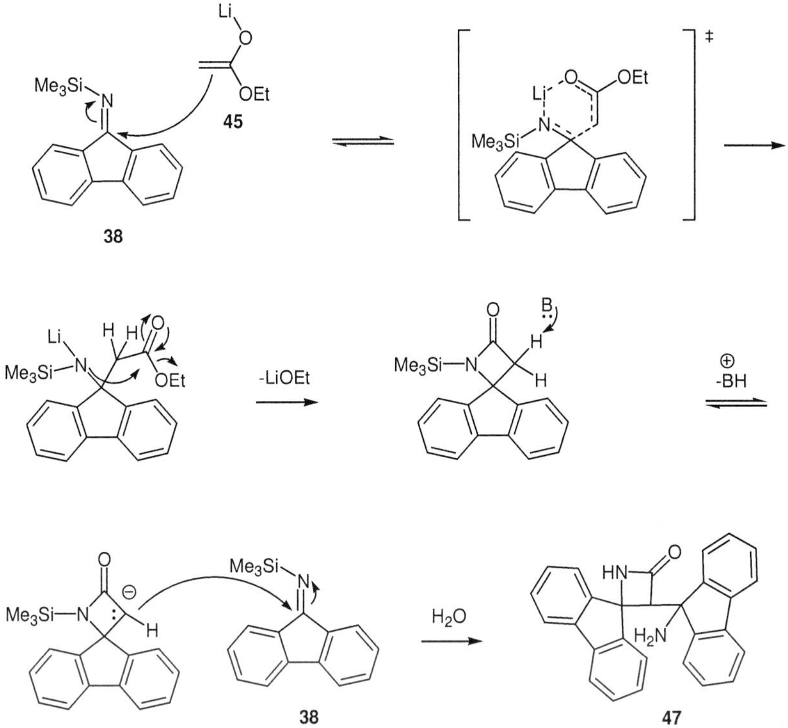

Table 16. Results of the reactions between fluorenone N-trimethylsilylimine **38** and Lithium enolate **45**

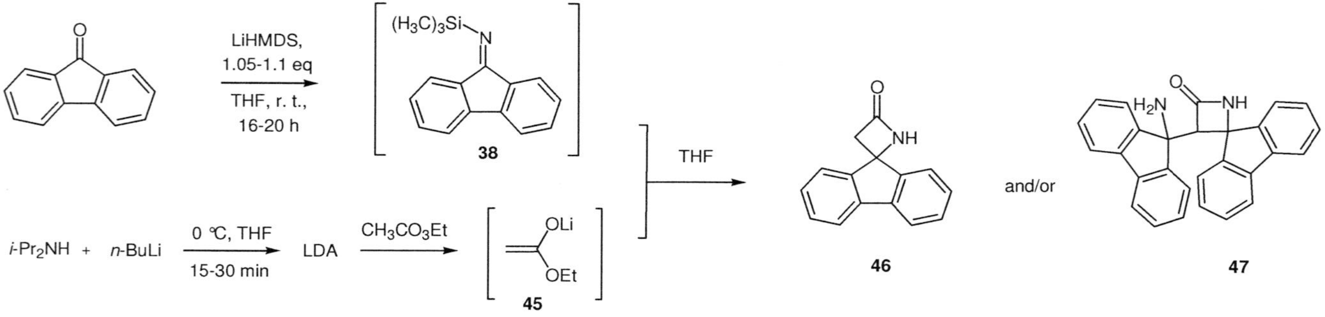

Entry	(i-Pr)$_2$NH [equiv.]	n-BuLi, [equiv.]	CH$_3$CO$_2$Et, [equiv.]	45 [equiv.]	Reaction conditions	Product(s), yield [%]	Experiment number
1	1.50	1.50	1.65	1.5	-78 °C to r.t., 16 h	**46** (11) and fluorenone (32)	47
2	3.31	3.31	3.00	3.0	0 °C to r.t., 20 h	**46** (40)	48
3	2.96	2.96	2.00	2.0	0 °C to r.t., 17 h	**46** (34) and **47** (25)	49
4	3.31	3.23	3.00	3.0	-60 °C to 0 °C, 17 h	**47** (53)	50
5	1.65	1.50	2.25	1.5	-78 °C to r.t., 2 d	fluorenone (39) and fluorenimine (40)	–

4.2.3 Application of *N-p*-methoxyphenyl-substituted imines as a ketimine substrate

The results of the experiments with fluorenone *N*-trimethylsilylimine (**38**) showed that different reaction conditions were necessary in order to obtain reliable results. The application of another imine-protecting group might be one possibility to prepare the required 3-unsubstituted β-lactams. The easily available *N-p*-methoxyphenyl imines **48** and **49** were the next substrates of choice. They were treated with ca. 1.5 equiv. of the lithium enolate of ethyl acetate (Scheme 27). The reactions were carried out in the temperature range −78 °C to +50 °C. However, only starting material was then reisolated.

Scheme 27. Attempted reactions of N-p-methoxyphenyl imines **48** *and* **49** *with the lithium enolate of ethyl acetate (**45**)*

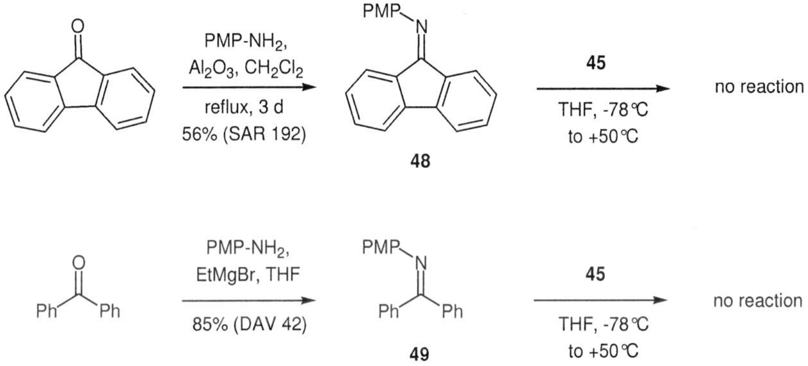

4.2.4 Conclusion and outlook

As shown by the results in Table 16, the reaction between the *N*-trimethylsilylimine **38** and lithium enolate **45** proved to be unreproducible. The interesting outcome of these experiments is the substitution at the C-3 atom of the β-lactam ring and the formation of the "dimer" **47**. This case is known in the chemistry of β-lactams.[118] It would be an interesting way of introduction of voluminous moieties into the C-3 position of the β-lactam **46**, when reliable conditions for this type of substitution can found.

The application of the *p*-methoxyphenyl protecting group for the fluorenone and benzophenone imines proved not to be effective on the way to the 4-unsubstituted β-lactams.

tert-Butyloxycarbonyl protection might be a promising alternative for such a ketimine in the reactions with lithium enolates.

So far, the approach described did not prove to be suitable for the preparation of 4-unsubstituted β-lactams. In order to open an access to the target β-amino acids, the well-know Staudinger reaction had to be applied. These results are presented in the following chapter.

4.3 The classical way to β-lactams – Staudinger reaction

4.3.1 Background

Staudinger et al. were the first to prepare a β-lactam by coupling a ketene, generated from an acid chloride, with an imine. The ketene can be generated, for example, thermally, photochemically, or by the reaction of acid chlorides and tertiary amines. This proceeds either through the abstraction of the α-standing proton and elimination of the chloride ion, or through the formation of acetyltriethylammonium ion with following elimination of triethylammonium chloride. The next step is nucleophilic addition of the imine to the ketene that results in the formation of the zwitterion intermediate. The following cycloaddition leads to the 2-azetidinone heterocycle (Scheme 28).[17,123]

Recently, many asymmetric variants of the Staudinger reaction have been presented.[17,124] In the chemical literature, this reaction is considered as a [2+2] cycloaddition of ketenes with imines. Usually, monosubstituted ketenes are applied, and reactions of disubstituted ketenes are much less common. In the present work, it was first of all important to develop the conditions of the reaction between a ketimine, derived from the symmetric ketones benzophenone and fluorenone, and a corresponding ketene. Afterwards, the conditions could be extended to ketimines deriving from unsymmetric ketones such as 1-indanone or 1-tetralone.

The target compounds are β-amino acids without substituents in the α-position. Therefore, the C-3 atom of the β-lactam should be unsubstituted as well. In order to fulfil this condition, acetyl chloride was chosen to generate the corresponding ketene. Unfortunately, its reaction with an *N-p*-methoxyphenyl-substituted imine and triethylamine as a base failed to give the required azetidinone structure.[125]

Scheme 28. Mechanism of the Staudinger reaction[123]

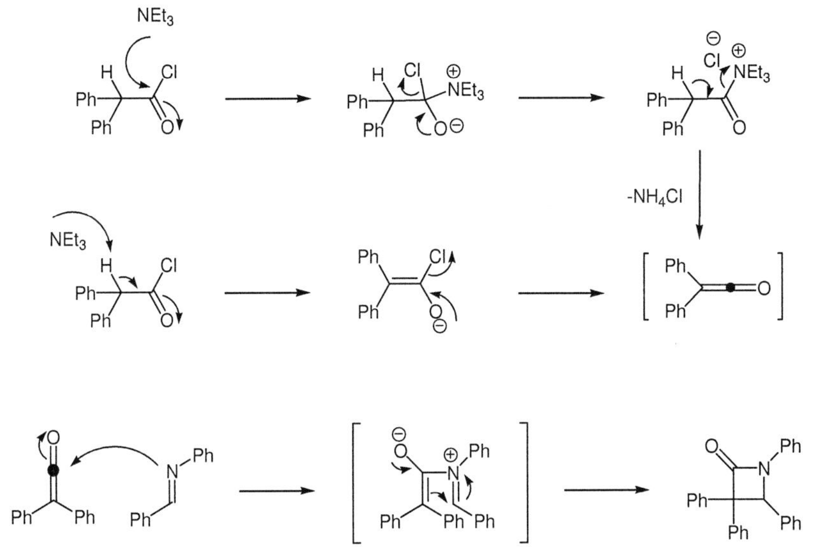

4.3.2 Own results

Quite good results were obtained by applying chloroacetyl chloride for the preparation of the corresponding ketene which reacted with the *N-p*-methoxyphenyl imines **48** and **49** giving the 3-chloro-substituted 2-azetidinones **50** and **52** in good yields (Schemes 29 and 30).

The chlorine atom at position 3 of the β-lactam ring is considered a disadvantage of this route to α-unsubstituted β-lactams. Fortunately, this substituent can easily be removed in a radical reaction in the presence of *tris*(trimethylsilyl)silane and AIBN as a radical starter.[126] The yields of *N*-protected 2-azetidinones **51** and **53** were 86 and 95 %, respectively.

The following step was to remove the *N-p*-methoxyphenyl protection. This procedure is known to run smoothly in the presence of an excess of ceric ammonium nitrate and at ambient temperature.[101,102] Application of the standard method to the *N*-PMP-lactams **51** and **53** led to low yields of the expected products due to decomposition of both starting material and product. Since this method is known to be applicable to substrates containing the β-lactam ring, the decomposition of the starting material might therefore have taken place due to the aromatic substituents at C-4 of **51** and **53**. In order to avoid this, the reaction was carried out at −10 °C followed by rapid work-up. The excess of ceric ammonium nitrate was destroyed by washing the reaction mixture with a solution of sodium bisulfite. Still, the yields of β-lactams **46** and **54** (43 % and 27 % over 3 steps, respectively) were not satisfactory.

The excess of ceric ammonium nitrate (CAN) was essential for the reaction to be complete. When only 1.5 equivalents of CAN were used at −10 °C, even after 20 h starting material was detected in the reaction mixture.

*Scheme 29. Synthesis of the β-lactam **46** by means of the Staudinger reaction*

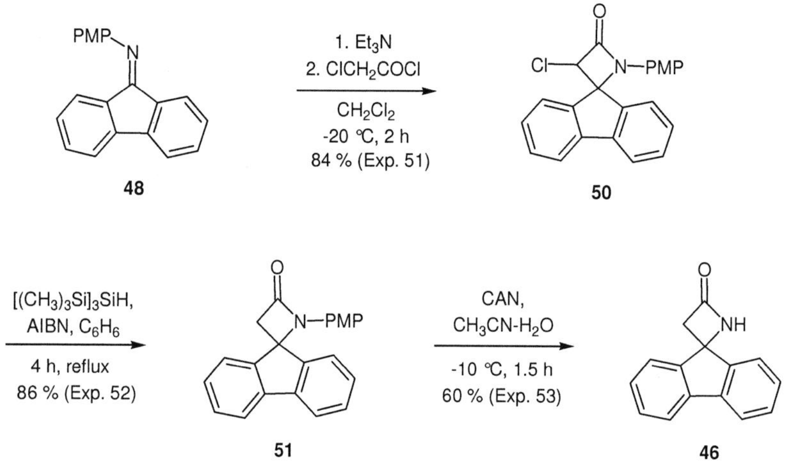

*Scheme 30. Synthesis of the β-lactam **54** by means of the Staudinger reaction*

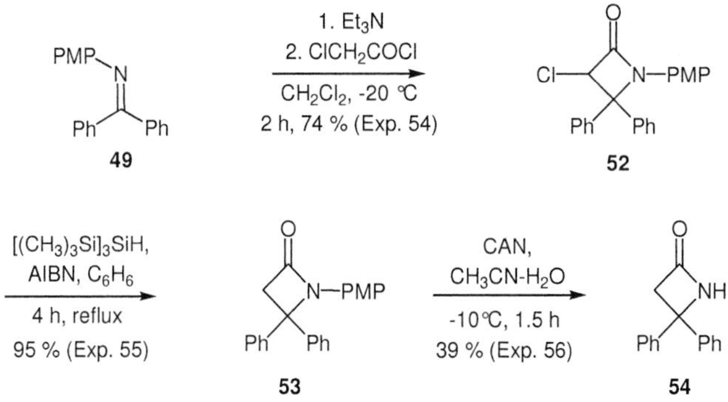

The structure of the ß-lactam **54** was confirmed by X-ray crystal structure analysis (Fig. 8).

*Figure 8. Crystal structure of the β-lactam **54** by X-ray diffraction analysis*

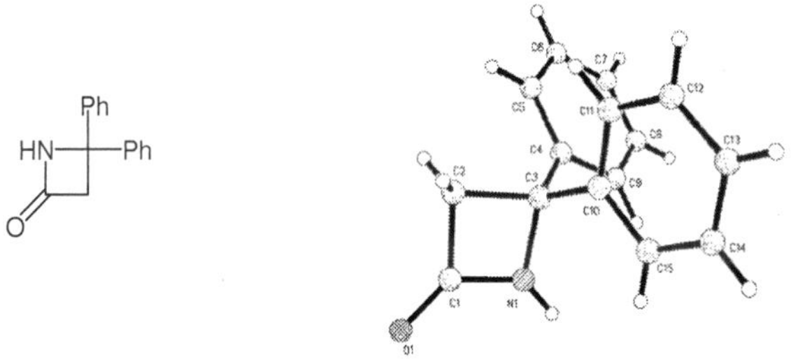

4.3.3 Application of the fluorenone *N*-trimethylsilylimine (38) as an imine component

In order to avoid problems related to the cleavage of the PMP protecting group, the *N*-trimethylsilyl-protected imine **38** was used (Scheme 31). The treatment with chloroacetyl chloride gave the 3-chlorosubstituted β-lactam **55**, with the released amino group. However, the yield of the 3-chlorosubstituted β-lactam **55** (59 %) was not good enough, so this could not be a valuable alternative to the approaches shown in Schemes 29 and 30 unless the reaction conditions were optimized.

Scheme 31. Change of the imine component in the Staudinger reaction

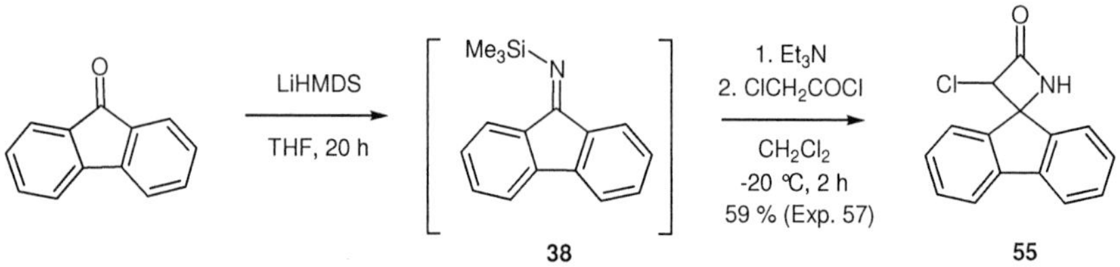

4.3.4 Conclusion and outlook

In conclusion, the Staudinger reaction was shown to be a reliable way to the target β-amino acids **E-I** *via* the corresponding β-lactams. For the present results, the methods for the ring-opening have to be investigated. Preliminary experiments on the acidic hydrolysis[127] of the 2-azetidinones **45** and **53** did not show good results, only decomposition was observed.

The cleavage of the PMP group should also be optimized. An alternative, mild procedure for its removal was recently proposed in the literature: The combination of periodic acid and trichloroisocyanuric acid (TCCA) was found to be particularly effective in liberation of the amino group.[128]

The problem of the deprotection might also be solved by application of the *N*-Boc-imines of fluorenone and benzophenone.

The Staudinger reaction can also be applied for the synthesis of β-amino acids with two different alkyl substituents in the β-position (compounds **G-I**). In order to reach enantiomerically pure products, asymmetric variants of this reaction have to be developed.

4.4 Discussion of the NMR spectra of fluorene-containing compounds

In this dissertation, a series of fluorene-containing compounds was synthesized. The majority of these molecules had a mirror plane, therefore their ^{13}C NMR spectra as expected showed the same chemical shifts for the signals of the corresponding atoms belonging to the rings A and B (1 and 8, 2 and 7 and so on). For the β-lactam **46**, it was possible to assign all the ^{1}H and ^{13}C NMR signal by means of the HSQC and HMBC experiments (Figure 9).

Heteronuclear Single Quantum Coherence (HSQC) experiments correlate chemical shifts of directly bound nuclei. Heteronuclear Shift Correlations *via* Multiple Bond Connectivities (HMBC) experiments allow to correlate carbons (or other X nuclei) and protons linked over more than one bond.[129]

The same pattern of the ^{13}C signals of the compound **46** could be also observed for other symmetric molecules containing the fluorene residue (for example **37-42, 51, 57**). Therefore the assignment of their NMR spectra was possible using the known ^{1}H and ^{13}C spectra of the β-lactam **46**.

*Figure 9. ^{1}H and ^{13}C NMR shifts (ppm) of the 2-azetidinone **46***

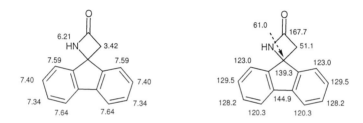

The signals of the protons at C-1(8) and C-4(5) in ^{1}H NMR spectra of the symmetric molecules can approximately be described as doublets with coupling constants *J* equal to 7.2-7.5 Hz. The protons at C-2(7) and C-3(6) gave signals in form of "triplets" with coupling constants ranging between 7.2 and 7.5 Hz. This observation is also in accordance with a literature data for an ABCD spin system of 1,2-disubstituted benzene derivatives.[129] These signals appeared often as multiplets where the coupling constants could not be determined.

Each atom of a pair (1-8, 2-7, 10-11 and so on) of some asymmetric compounds, such as **36, 47, 50** or **55**, gave an individual signal in the ^{13}C spectra. This led to a double set of signals in such spectra, therefore they became rather complicated. It was not possible to assign each individual signal with certainty based just on the ^{13}C spectra of the compound **46**. The assignment might be possible applying special measurements such as HSQC or HMBC experiments; however they were not done within the frame of this dissertation. The HSQC and HMBC spectra were recorded for the 3,3,4-trisubstituted 2-azetidinone **47**, having two fluorenyl residues, but it was not possible to assign the individual signals because of the complexity of the spectra (see Appendix).

5   APPLICATION OF THE LITHIUM ENOLATE OF *N,N*-DIMETHYLACETAMIDE

5.1   Literature survey

In Chapter 4.2 the reaction between the fluorenone *N*-trimethylsilylimine (**38**) and the lithium enolate of ethyl acetate (**45**) was discussed. As Table 16 shows, the results were not reproducible and difficult to interpret. One of the possible problems could be the behaviour of the lithium enolate **45**. Though its application is known in the literature,[117,130] only poor yields are reported. To scan this aspect, *N,N*-dimethylacetamide was taken as an ester component. If the lithium enolate of this acetamide would be used, the cyclization of the β-amino ester formed should be blocked (see mechanism shown in Scheme 24).

Mukaiyama *et al.*[131] reported on the synthesis of *N*-acetyl-L-daunosamine (**AX**) which, having a free amino group, is an important part of some antibiotics. In this total sequence, the key step was a stereoselective carbon-carbon bond forming reaction between α-lithio *N,N*-dimethylacetamide and the imine **AV** (Scheme 32). Zink halide as a Lewis acid provided the selectivity of the addition to the imine **AV**. The β-amino amide **AW** was the "predominant product"; no data concerning the exact diastereomeric ratio of the addition products **AW** were given, however.

*Scheme 32. Preparation of the N-acetyl-L-daunosamine (**AX**)[131]*

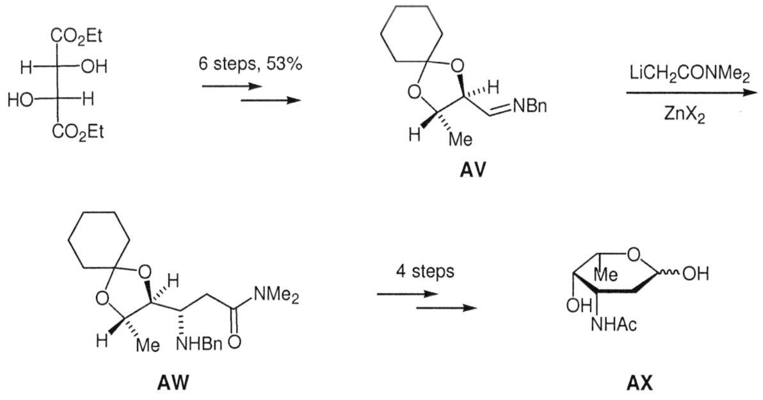

This example demonstrates the principal possibility of the introduction of a nucleophilic C_2-component with a terminal amino function.

The lithium enolate of *N,N*-dimethylacetamide (**56**) can be prepared according to a standard method from the corresponding acetamide by treatment with a solution of lithium diisopropylamide at 0 °C in THF (equation 5).[132]

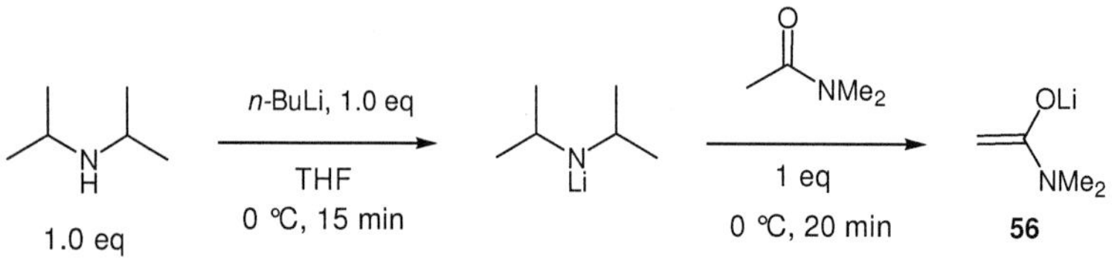

(5)

5.2 Own results

Previous experiments (Chapters 3.4, 4.2, 4.3) had shown that the preparation of the Fluorenone *N*-trimethylsilylimine (**38**) was quite easy and reliable. Thus, this compound was chosen as an imine component to find optimal conditions for the addition of the lithium enolate **56** to the C=N imine bond (Table 17).

*Table 17. Reactions between the lithium enolate **56** and the N-trimethylsilylimine **38***

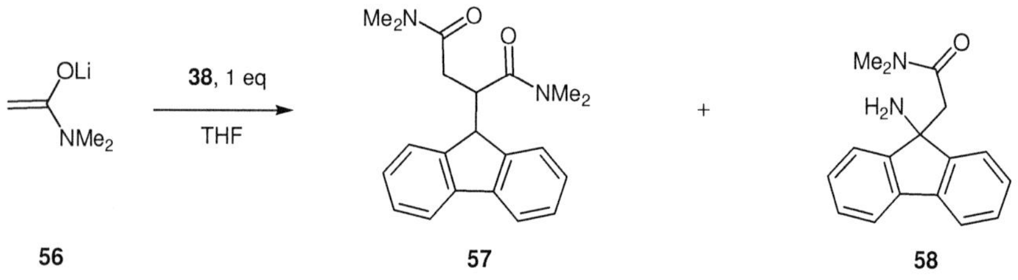

Entry	56 [equiv.]	Reaction conditions	Product(s), yield [%][a]		Experiment number
			57	**58**	
1	1.5	0 °C, 3 h; r.t., 2 h	27	"61"	58
2	1.2	–30 °C, 16 h	45	"15"	59
3	1.5	–50 °C, 7 h	traces[b]	77[b]	60
4	2.5	0 °C, 2 h	"138"[c]	–	–

[a]Yield based on the amount of fluorenone used. – [b]The amide **58** was isolated with traces of the diamide **57** ("23 %"). – [c]In a mixture with *N,N*-dimethylacetamide.

In the first experiment (Table 17, entry 1), the *N*-trimethylsilylimine **38** was treated with 1.5 equivalents of the lithium enolate **56**. The reaction was run for 3 h at 0 °C and then for 2 h at ambient temperature, followed by quenching with water. After trituration of the crude product in diethyl ether, a colourless solid of **57** formed and was isolated in 27 % yield (based on the amount of the fluorenone used). This diamide product was not the required β-amino amide **58**; the structure could be established on the base of X-ray crystal structure analysis as that of the "2:1" adduct, as shown in the Figure 10.

*Figure 10. X-ray diffraction analysis of crystal structure of the dimeric product **57***

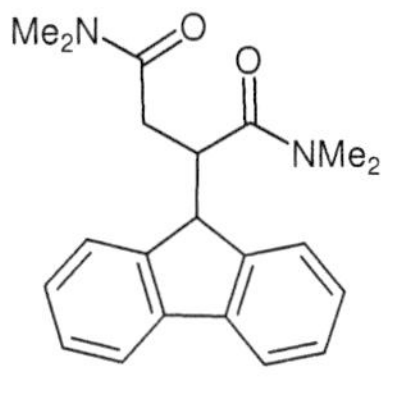

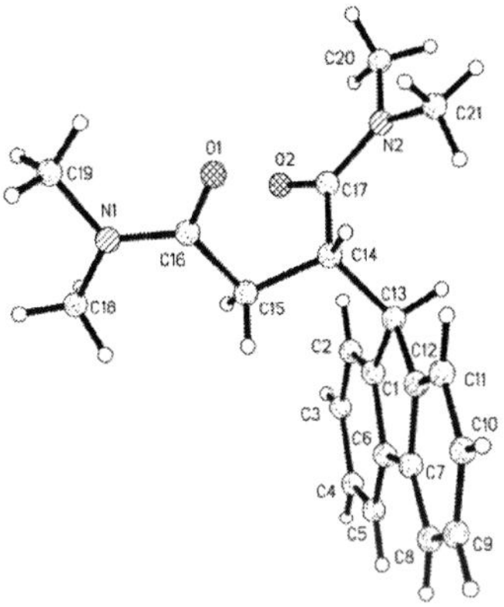

The "normal" product, the β-amino amide **58**, was left in the mother liquid and was not isolated in this experiment. The comparison of the NMR data with those of the following experiments proved that the yield of the crude monoamide **58** was ca. 61 %.

The reason for the formation of the succinic product **57** was not clear. Assuming that the excess of the lithium enolate **56** promotes the generation of this by-product, the reaction was repeated with 1.2 equivalents of **56** and at –30 °C (Table 17, entry 2). In this case, the major product (45 %) was the "dimeric" one (**57**). The β-amino amide **58** was also not purified in this experiment. Later, its yield was established to be around 15 %. So far, the assumption that the excess of the lithium enolate **56** provoked the formation of the diamide product **57** was not confirmed.

The experiment was performed again with 1.5 equivalents of the lithium enolate **56**, at –50 °C (Table 17, entry 3). The analysis of the crude product showed that the β-amino amide **58** was the major component in the mixture. It was purified and isolated in 77 % yield as a sticky oil. The residue ("23 %") consisted of the crude β-amino amide **58** with only traces of the 2:1 adduct **57** (according to NMR data).

When a large excess of the lithium enolate of *N,N*-dimethylacetamide (**56**) was used at 0 °C (Table 17, entry 4), exclusively the succinic product **57** formed after 2 h reaction time (as NMR data showed). This meant that the excess of the enolate accelerated the formation of the succinic diamide dramatically, as comparison with entry 1 of Table 17 shows.

The analysis of the data (Table 17, entries 1-3) showed that the yield of the succinic diamide **57** did not depend strongly neither on the amount of the lithium enolate used in the reaction nor on the temperature. Probably, this diamide product is generated from the β-amino amide **58** during the reaction. The longer the reaction time, the more **57** will be formed. It is also possible that, at comparable reaction times, the lower the temperature the slower is the formation of **57** (compare entries 1 and 3). This assumption was confirmed by the fact that the use of a large excess of the lithium enolate of ethyl acetate (**56**) did not lead to the formation of the required β-amino amide **58**.

Thus, this reaction demands only a slight excess of the enolate, possibly low temperature and short reaction time only necessary for the formation of the required product **58**. The conditions illustrated by entry 3 so far are the best ones for isolation of the β-amino amide **58**.

5.2.1 Mechanism of the formation of the succinic derivative 57

Before one starts considering the possible mechanism of the formation of **57**, one has to keep in mind that the fluorenylmethoxycarbonyl group (Fmoc) is widely used in peptide synthesis for the protection of the amino function. This group is stable under acidic conditions, but can easily be removed under very mild basic conditions (*e.g.* piperidine in DMF).[133] The mechanism of this process is shown in Scheme 33. The reason why the Fmoc group is not stable in the presence of base is the possibility of delocalization of the negative charge due to the formation of aromatic fluorenyl anion and concomitant case of elimination.

Scheme 33. Mechanism of the Fmoc deprotection[63]

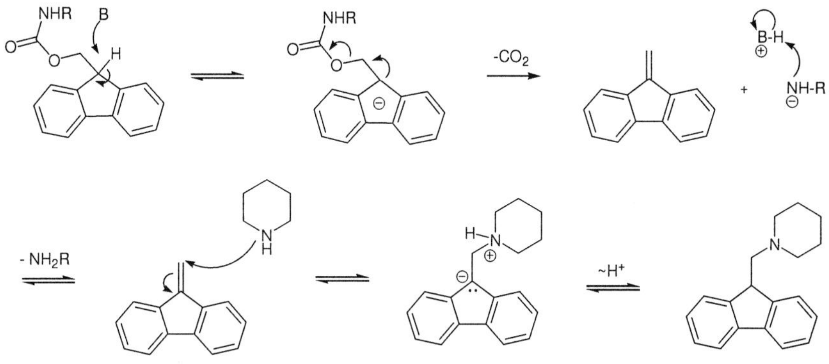

The same process might be important in the course of the reaction between the lithium enolate **56** and fluorenone *N*-trimethylsilylimine **38**. The following mechanism is proposed:

The reaction starts by nucleophilic attack of the enolate **56** to the electrophilic C-9 position of the fluorene in the *N*-trimethylsilylimine **38** followed by C-C bond formation (Scheme 34). The intermediate **AY** is a precursor of the β-amino-*N,N*-dimethylamide **58**, and can be considered as the kinetic product of the reaction. If the reaction time is long enough, the intermediate **AY** eliminates the amino function with the formation of the conjugated 9-fluorenylmethylene derivative **AZ**. The latter can be attacked by the lithium enolate **56** to give, after hydrolysis, the succinic diamide **57**. The intermediate **BA** is the thermodynamic product of this reaction.

*Scheme 34. Proposed mechanism of the formation of the 2:1 adduct **57***

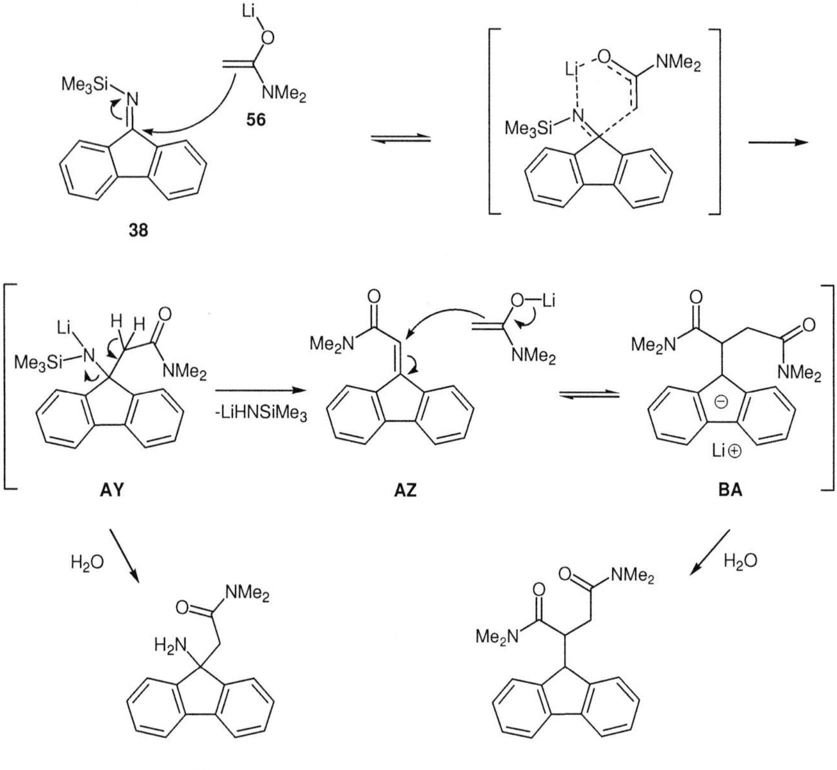

5.2.2 Hydrolysis of the *N,N*-dimethylamide group

Knowing the exact course of the reaction (Scheme 34), it is possible to obtain only the required product, in this case the compound **58**. It is a direct precursor of 9-amino-9-fluorenylacetic acid (**E**). The only step remaining would be the hydrolysis of the amide group releasing the carboxy group. The simplest method known in the literature is acidic hydrolysis,[63] but in the case of the *N,N*-dimethylamide **58** there was a risk of decomposition (compare with the acidic hydrolysis of the β-lactam **46** (Chapter 4.3). Milder conditions have been suggested by Mukaiyama *et al.*,[131] namely treatment with 80 % aqueous acetic acid under reflux. However, heating of *N,N*-dimethylacetamide **58** under these conditions for 25 h did not change the starting material. Afterwards, ca. 3.4 equivalents of hydrochloric acid were added to the solution and it was heated for additional 20 h. This procedure did not result in the hydrolysis of the amido group but in the formation of the hydrochloric salt of the starting material (**58**·HCl, Exp. 61). Thus, the acidic hydrolysis of the amide function failed to give the corresponding free amino acid.

Amide hydrolysis by aqueous base is also known. Since the β-amino-*N,N*-dimetylacetamide derivative **58** represents a bulky tertiary amide, it was promising to apply the conditions described by Gassman *et al.*[134] In this approach, the corresponding amide is treated with an excess (ca. 6 equivalents) of potassium *tert*-butoxide in the presence of water (2 equivalents). The reaction of water with butoxide generates potassium hydroxide and therefore an anhydrous medium. The strongly nucleophilic hydroxide adds to the carbonyl group of the amide **BB** to produce the anion **BC**. The conversion of **BC** into the dianion **BD** is accomplished by the excess of potassium *tert*-butoxide followed by the cleavage of the dimethylamide anion to form the salt of the corresponding acid (Scheme 35).

Scheme 35. Basic hydrolysis of bulky tertiary amides[134]

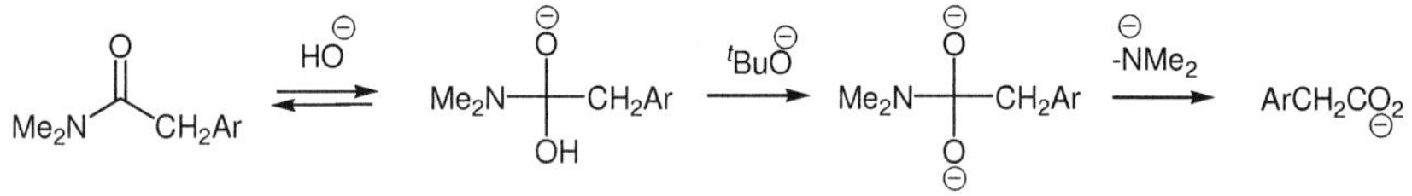

The Gassman reaction conditions were applied to the hydrolysis of the *N,N*-dimethylamide **58**. In order to ease the isolation of the corresponding carboxylic acid, the product **58** was converted to the corresponding *N*-Boc derivative **59** (Exp. 62). The latter was subjected to basic hydrolysis as illustrated by Scheme 34. The reaction was left with stirring for 3 days at ambient temperature, however, only starting material was detected in the mixture. Following heating for 3 hours at 65 °C did not lead to the consumption of the starting tertiar amide **58**. Eventually, it was reisolated.

5.2.3 Conclusion and outlook

In summary, the reaction between the fluorenone *N*-trimethylsilylimine (**38**) and the lithium enolate of the *N,N*-dimethylacetamide (**56**) proved to be a promising route to the corresponding β-amino *N,N*-dimethylacetamide **58**. This reaction did not give side reactions under well-defined conditions; no starting material was detectable. The only problem so far is the hydrolysis of the tertiary amide under acid or basic conditions, which was not yet successful.

The problem with the hydrolysis might be avoided by application of acetonitrile instead of the *N,N*-dimethylacetamide (Scheme 36). Acetonitrile has three C–H acidic protons one of which can be abstracted in the presence of a strong base such as LDA. The corresponding lithiated derivative could then attack the *N*-trimethylsilyl imine **38** as a nucleophile.[135] The advantage here would consists in the relative simplicity of the hydrolysis of the nitrile group.

Scheme 36. Potential application of lithiated acetonitrile

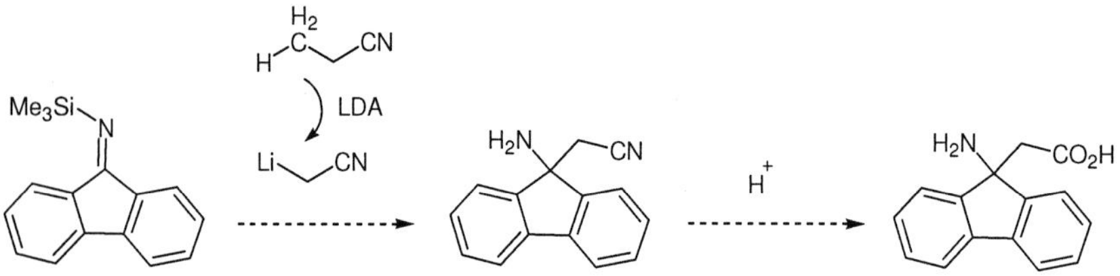

6   DETAILED SUMMARY AND OUTLOOK

The interest in asymmetric synthesis of α- and β-amino acids with voluminous substituents originates from their special properties and applications in chemical and medicinal research. In the present work, 1-adamantylglycine **D** (α-amino acid) as well as conformationally constrained β-amino acids **E–I** were chosen as attractive target structures.

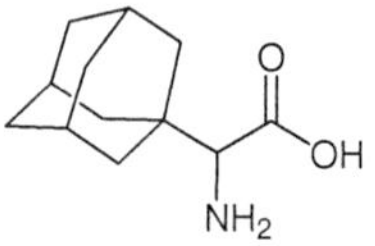

D	**E**	**F**

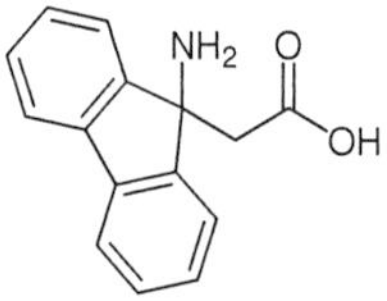

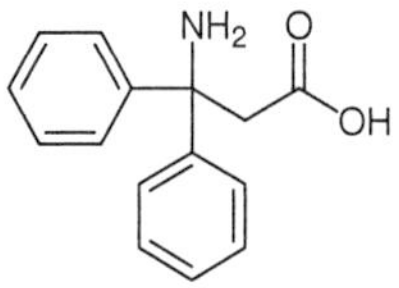

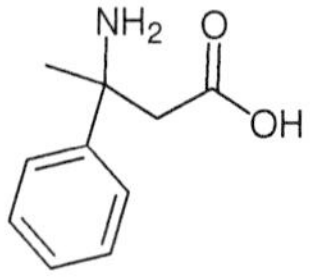

G	**H**	**I**

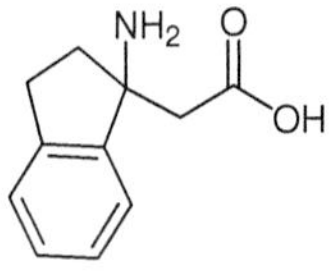

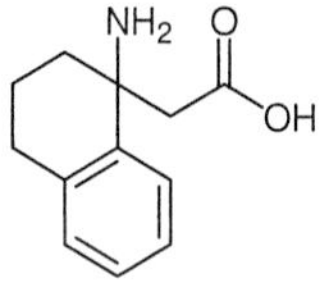

The following routes to α- and β-amino acids were proposed:

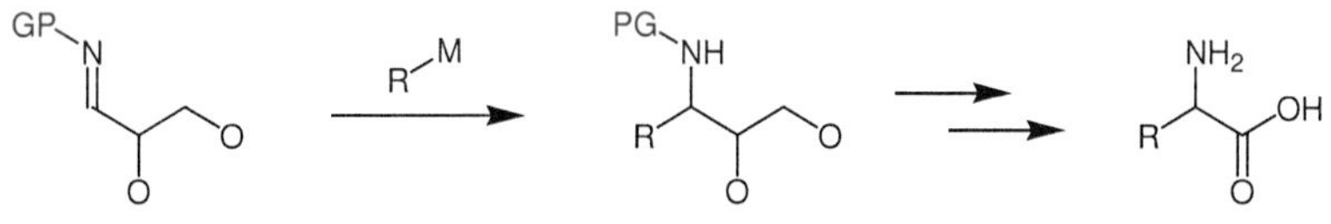

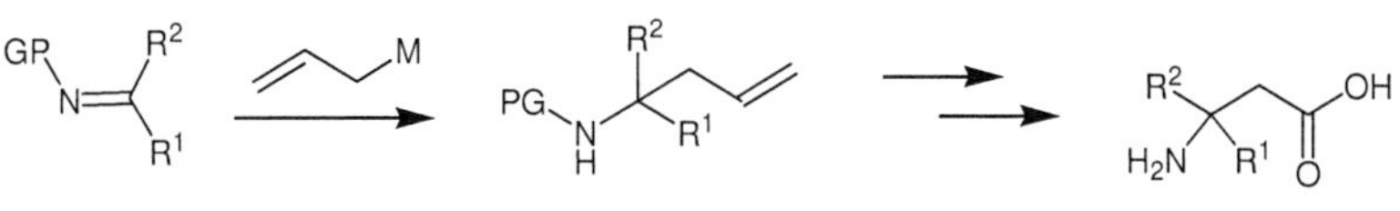

R, R^1, R^2 = alkyl/aryl rests; M = MgBr, MgCl or Li; PG = protecting group

They include the addition of various Grignard reagents to imines or nitrones with suitable *N*-protecting/activating groups, followed by oxidative transformation of the allyl moiety. These pathways would open an access to a structural variety of amino acids. At the same time, they might allow for control of the stereoselectivity of the nucleophilic addition.

A) Synthesis of 1-adamantylglycine

A series of reactions concerning additions to imines or related C=N compounds was carried out in this work with phenyl, *tert*-butyl, and 1-adamantyl Grignard reagents. Instead of an imine, 2,3-*O*-isopropylidene-D-glyceraldehyde *N*-benzylnitrone (**4**) was applied since nitrones are known to be more stable as compared to structurally similar glyceraldehyde imines. The results obtained are summarized in Table 7 (eq. 5).

| **4** | *threo (syn)* | *erythro (anti)* | (5) |

The success of the addition of 1-adamantylmagnesium- and phenylmagnesium bromides to the nitrone **4** was shown to depend on the presence or absence of a Lewis acid. The addition of adamantylmagnesium bromide to the *N*-benzylnitrone (**4**) without precomplexing with a Lewis acid was non-selective; the presence of Et_2AlCl led to an increase of the diastereomeric ratio *threo:erythro* up to <5 : 95. The presence of $ZnBr_2$ changed this ratio to 58:42 (Table 7, entries 1-3).

The attempts to separate the diastereomeric *N*-hydroxylamines derived from **4** were successful only in the case of R = Ph. The configurations of the *N*-hydroxylamines with R = *t*-Bu were not established. The configuration of the *N*-hydroxylamine **7b** (R = Ad) could be determined after transformations into adamantane-containing products.

For the preparation of 1-adamantylglycine, the 2,3-*O*-cyclohexylidene-D-glyceraldehyde nitrone **16** was used, where the cyclohexylidene protecting group is more stable with respect to acids as compared to the isopropylidene protection of the *N*-benzylnitrone **4** (Scheme 11). In fact, when the cyclohexylidene protecting group was used instead of isopropylidene, the yield of *N*-benzylhydroxylamine **17** was improved to 61 % (compare with the 42 % yield for the *N*-benzylhydroxylamine **7b**, entry 8, Table 7). The configuration (*erythro*) of **17** was determined later, after conversion into already known (*S*)-1-adamantylglycine hydrochloride [(*S*)-**24**].

Table 7. Addition of some organomagnesium and -lithium reagents to 2,3-O-isopropylidene-D-glyceraldehyde N-benzylnitrone (4)

Entry	RM	Equiv.	Lewis acid	Time	threo : erythro[c]	Yield, mixture [%]	Yield of isolated diastereomers [%]	Experim. number
1	PhMgBr	1.5	ZnBr$_2$	6 h	58 : 42	48	–	1
2	PhMgBr	1.5	none	6 h	75 : 25	82	33 (5a) + 13 (5b)	2
3	PhMgBr	1.5	Et$_2$AlCl	16 h	<5 : 95	91	70	5
4	PhLi	3	none	40 min	85 : 15	73	46 (5a) + 9 (5b)	3
5	t-BuLi	1.5	none	40 min	75 : 25[d]	20	–	6
6	t-BuMgCl	2	none	6 h	78 : 22[d]	76	12 (6a or 6b) 9 (6a+6b)	7
7	AdMgBr	2	none	17 h	42 : 58	51	–	8
8	AdMgBr	2	Et$_2$AlCl	16 h	<5 : 95	42	–	9

[a]Solvent used: diethyl ether, except of entry 2 (THF). - [b]Temperature: −60 °C for M = MgBr and −85 °C for M = Li. - [c]Measured from the intensities of ^{13}C NMR signals. - [d]The configurations of the stereoisomers have not been established.

The addition product **17** was hydrogenated in the presence of Pearlman's catalyst. The corresponding amine **19** was obtained in 87 % yield and then treated with 12 N HCl to release the 3-amino-1,2-diol hydrochloride **21** (82 % yield). This hydrochloride was then treated with di-*tert*-butyl dicarbonate and gave the corresponding *N*-protected aminodiol **22** in 79 % yield (Scheme 11).

The transformation of the diol moiety was done by oxidative cleavage with sodium periodate followed by oxidation to the carboxylic acid with sodium chlorite. This procedure afforded *N*-Boc-1-adamantylglycine [(*S*)-**23**] in good yield (82 %). The following cleavage of the *N*-Boc protection gave (*S*)-1-adamantylglycine hydrochloride [(*S*)-**24**]. Thus, the yield of the target amino acid was 28 % over 6 steps.

*Scheme 11. Synthesis of (S)-1-adamantylglycine hydrochloride [(S)-**24**]*

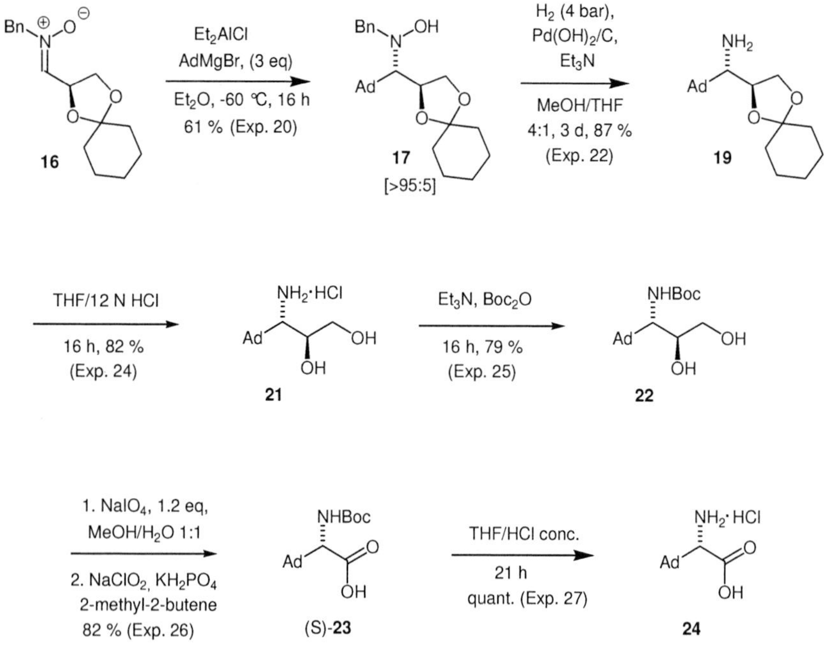

B) Preparation of β-amino-β,β-diphenylpropionic acid (F)

Benzophenone has served as a starting material (Scheme 16). In order to prepare the protected imine **25**, benzophenone and *p*-methoxybenzyl amine were heated to reflux in toluene in the presence of $ZnBr_2$ as a Lewis acid with azeotropic destillation of water. The treatment of the **25** with allylmagnesium bromide took place smoothly and gave the allyl derivative **26** in excellent yield.

For the preparation of the diol **27** two methods have been used (A, B, Scheme 16). The reaction with osmium tetroxide in the presence of *N*-methylmorpholine *N*-oxide as reoxidizing agent[136] afforded the required compound in only 39 % yield, which was not satisfactory. The diol **27** was then obtained in better yield (74 %) using potassium osmate dihydrate as a catalyst and potassium hexacyanoferrate as reoxidant. The second method proceeded quite slowly and needed five days to be complete, as compared to 6 h with method A.

Scheme 16. Synthesis of methyl ester of N-protected β-amino-β,β-diphenylpropionic acid

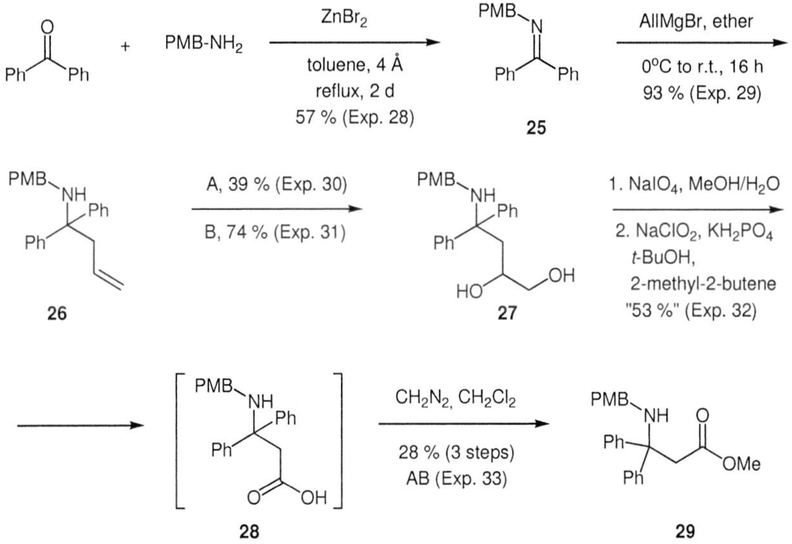

Method A: 1. 10 wt % NMO/H_2O, 2.5 wt % OsO_4/*t*-BuOH, *t*-BuOH/THF, 6 h. - 2. Na_2SO_3
- 3. Column chromatography.
Method B: 1. $K_3Fe(CN)_6$, $K_2OsO_2(OH)_4$, K_2CO_3, *t*-BuOH/H_2O, 5 d. -
2. Na_2SO_3. - 3. Recrystallization

The diol **27** was oxidatively transformed to the 3-(4-methoxybenzylamino)-3,3-diphenylpropanoic acid (**28**). The structure of this compound was not confirmed at this stage.

The NMR spectra could not be recorded since the product **28** is soluble neither in water nor in an organic NMR solvent. In order to overcome the problem, the *in situ* esterification of the acid **28** without isolation was carried out (Scheme 16). The methyl ester **29** was isolated without any problem, however with a quite low yield, 28 % over 3 steps starting from the diol **27**.

The compound **29** in fact constitutes the required amino acid **F** with both amino and carboxy functions protected. Therefore the optimization of the esterification step remains necessary.

C) Synthesis of 9-aminofluorenylacetic acid (E)

The preparartion of the amino acid **E**, chosen as a target, was planned according to already known sequence: Introduction of the allyl moiety by means of nucleophilic addition to the imine C=N bond with consequent oxidative degradation of the double bond of the olefin.

As illustrated in Scheme 22, fluorenone was treated with lithium bis(trimethylsilyl) amide to form the corresponding *N*-trimethylsilylimine **38** which was not isolated but directly treated with ethereal allylmagnesium bromide. The expected 9-allyl-9-aminofluorene (**39**) was isolated in 24 % yield only.

A much better yield was obtained when the crude amine **39** was isolated and, without purification, treated with one equivalent of *tert*-butyl dicarbonate in acetonitrile. On this way, the yield of **40** was 61 % over 3 steps.

*Scheme 22. Synthesis of 9-(N-Boc-amino)-fluorenylacetic acid (**42**)*

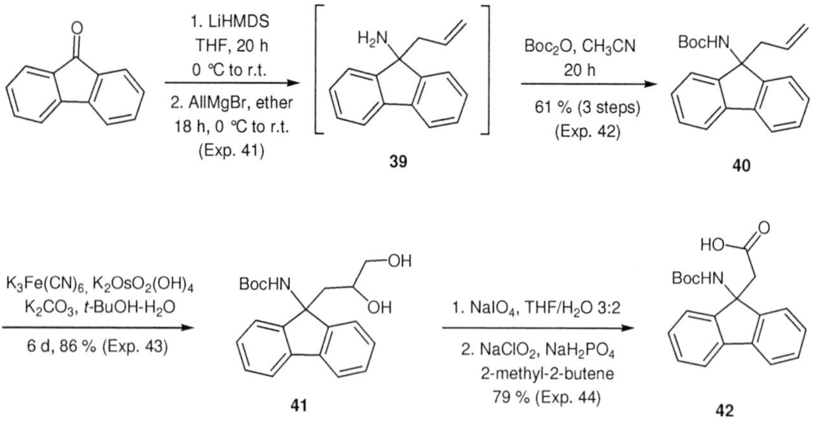

The double bond of the allyl moiety in compound **40** was derivatized in the presence of a catalytic amount of potassium osmate dihydrate and potassium hexacyanoferrate (Scheme 16). The transformation of the 1,2-diol moiety followed the known oxidative cleavage with sodium periodate and the ensuing oxidation of the corresponding aldehyde in the presence of sodium chlorite. The *N*-Boc-protected β-amino acid **42** was thus synthesized in 41 % yield over 6 steps.

D) Attempts on the preparation of spiro[azetidine-2,9'-[9*H*]fluorene]-4-one (46)

Several experiments on the preparation of the β-lactam **46** by means of the reaction between the *N*-trimethylsilylimine **38** and lithium enolate **45** have been performed (Scheme 37). The required 2-azeditinone product was isolated either in low yield or in a mixture with the product of the substitution at the C-3 atom of the β-lactam ring (**47**). The results proved to be unreproducible.

tert-Butyloxycarbonyl protection instead of the *N*-trimethylsilyl group might be a promising alternative for a ketimine like **38** in the reactions with lithium enolates.

Scheme 37. R*eaction between fluorenone N-trimethylsilylimine **38** and Lithium enolate **45***

E) Staudinger reaction

Better results concerning the synthesis of the β-lactam **46** were obtained by means of the Staudinger reaction (Schemes 29 and 30). The reaction of chloroacetyl chloride and the *N-p*-methoxyphenyl imines **48** and **49** in the presence of triethylamine gave the 3-chloro substituted 2-azetidinones **50** and **52** in good yields.

The chlorine atom at the carbon atom C-3 was easily removed in a radical reaction by use of *tris*(trimethylsilyl)silane and AIBN as a radical starter. The yields of *N*-protected 2-azetidinones **51** and **53** were 86 and 95 %, respectively.

In the following step the *N-p*-methoxyphenyl protection was removed by ceric ammonium nitrate, however the yields of the β-lactams **46** and **54** were not satisfactory. The reason for the low yields could be the destruction of the products by the excess of CAN during the work-up.

The total yields of β-lactams **46** and **54** (43 % and 27 % over 3 steps, respectively) might be improved by using milder conditions for the removal of the *N-p*-methoxyphenyl protecting groups. These 2-azetidinones are very important compounds since they are precursors of the target β-amino acids **E** and **F**.

*Scheme 29. Synthesis of the β-lactam **46** by means of the Staudinger reaction*

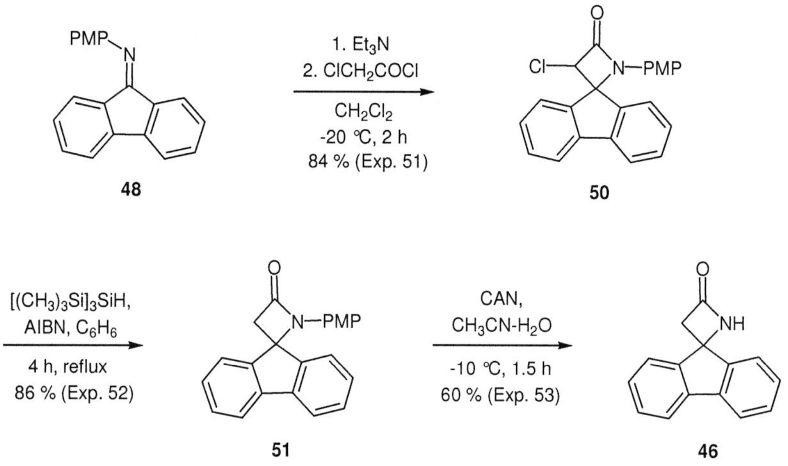

*Scheme 30. Synthesis of the β-lactam **54** by means of the Staudinger reaction*

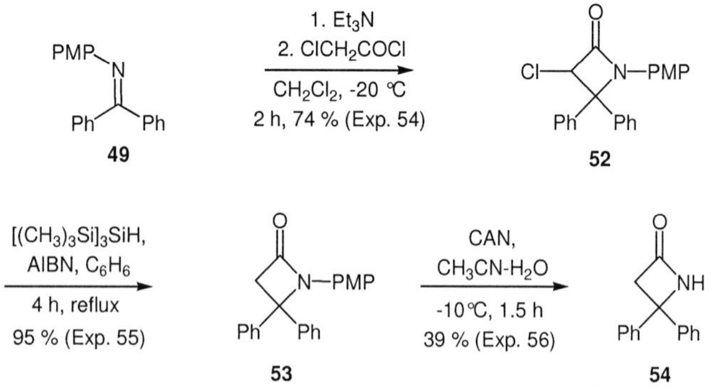

F) Application of the lithium enolate of *N,N*-dimethylacetamide (56)

The reactions between fluorenone *N*-trimethylsilylimine (**38**) and the lithium enolate **56** led to a mixture of the β-amino amide **58** and the succinic diamide **57** in various ratios, depending on the conditions (Table 17). The amide **58** is the precursor of the target amino acid **E**, and optimal conditions for its formation as a major product have been found (see entry 3 of Table 17). The *N,N*-dimethylamide group could not be hydrolyzed, neither in 80 % aqueous acetic acid nor with addition of the hydrochloric acid. An attempt of basic hydrolysis of the β-amino amide **58** (as its *N*-Boc derivative) in the presence of potassium *tert*-butoxide/potassium hydroxide mixture failed also and did not give the corresponding carboxylic acid.

*Table 17. Reactions between the Lithium enolate **56** and N-trimethylsilylimine **38***

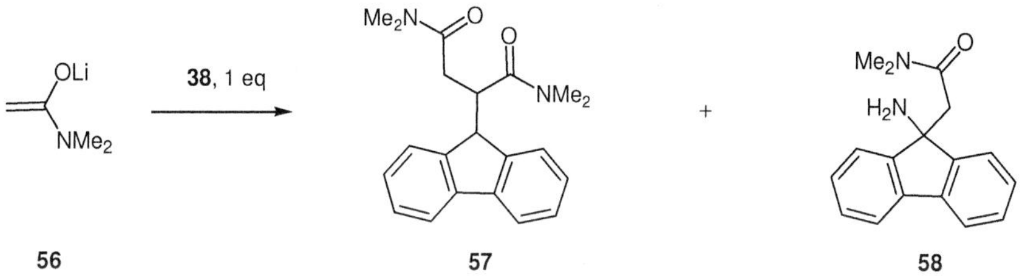

Entry	**56** [equiv.]	Reaction conditions	Product(s), yield [%][a]		Experiment number
			57	**58**	
1	1.5	0 °C, 3 h; r.t., 2 h	27	"61"	58
2	1.2	–30 °C, 16 h	45	"15"	59
3	1.5	–50 °C, 7 h	traces[b]	77[b]	60
4	2.5	0 °C, 2 h	"138"[c]	–	–

[a]Yield based on the amount of fluorenone used. – [b]Along with the main fraction, additional amount of the amide **58** was isolated ("23 %") with traces of the diamide **57**. – [c]A mixture with *N,N*-dimethylacetamide.

Outlook

The proposed way to (S)-1-adamantylglycine includes a highly diastereoselective addition step to a nitrone. The advantage consists in the control of the diastereoselectivity by choice of the Lewis acid applied. This means, both enantiomers of the substituted glycines are accessible by this route. This new route can easily be used for the preparation of substituted glycines with such bulky substituents as phenyl, *tert*-butyl and similar.

The approaches to the target β-amino-β,β-diphenylpropionic and 9-aminofluorenylacetic acids developed are based on imine additions and could provide an access to optically active β-amino acids (**G, H, I** and similar). In this case, asymmetric induction would be necessary (e.g. application of asymmetric catalysis or an *N*-chiral auxiliary).

7   EXPERIMENTAL PART

7.1   General

Nuclear Magnetic Resonance (NMR) spectroscopy

^{1}H NMR Spectra: Bruker AC 250 (250.1 MHz)
                       Bruker ARX 300 (300.1MHz)
                       Bruker ARX 500 (500.1 MHz)

^{13}C NMR Spectra: Bruker AC 250 (62.9 MHz)
                       Bruker ARX 300 (75.5MHz)
                       Bruker ARX 500 (125.8 MHz)

The chemical shifts are given in ppm. The TMS signal is taken as the reference ($\delta = 0.00$ ppm). The coupling constants (J) are given in Hertz (Hz). All chemical shift values and the multiplicity of NMR signals are shown with standard notations as follows: s (singlet), d (doublet), t (triplet), q (quartet), m (multiplet), b (broad signal).

The determination of the diastereomeric ratios is based on the intensities of separated signal pairs in the ^{13}C NMR spectra.[69] Assignment of the absorptions of carbon atoms has been done by means of C,H COSY spectroscopy.

Elemental analyses:

Were performed at the Institut für Organische Chemie, Universität Stuttgart.

Melting points:

Were measured with a Fisher-Johns heating apparatus and are not corrected.

Infrared spectroscopy:

FT-IR Spectra were recorded on a Bruker (IFS 28) spectrophotometer. The sample measurements were done directly without matrix. The position of the absorption bands are given in cm^{-1}, the intensities are given as follows: vs (very strong), s (strong), m (medium), w (weak), and b (broad).

Optical rotations

Angles of rotations were measured with the polarimeter 241 MC of Perkin-Elmer. The samples were dissolved in absolute solvent and filled into the cuvette. The optical rotations were calculated from the Na$_D$ absorption by means of the Drude equation by extrapolation of two Hg lines (546 and 578 nm):[137]

$$[\alpha]_D^T = \frac{[\alpha]_{578}^T \times 3.199}{4.199 - \frac{[\alpha]_{578}^T}{[\alpha]_{546}^T}} \qquad \text{with} \qquad [\alpha]_D^T = \frac{\alpha \times 100}{c \times d}$$

α = measured optical rotation value

c   = concentration, g/100 ml

d   = layer thickness in dm

λ = wavelenght, nm

T   = temperature, °C

Crystal structure analysis

For the X-ray crystal structure analysis a Nicolet P3 refractometer with graphite monochromator was used. The measurements were done with Mo-K$_\alpha$ wavelength. The calculation of the structures was done with SHELXS-86 or SHELXL-93[138], XRAY 76,[139] ORTEP II[140] and FRIEDA[141] programmes.

Thin layer chromatography (TLC):

Was performed on precoated aluminium sheets (silica gel 60 F254, layer thickness 0.2 mm) purchased from E. Merck. TLC was checked under UV light with wavelenght of 254 nm and/or stained with a solution prepared from 2 g KMnO$_4$, 20 g K$_2$CO$_3$, 5 mL of 5 % NaOH solution in 300 mL water and developed by means of a heat gun.[142]

Filtration and column chromatography

Silica gel 60 with mesh size 40-62 µm (E. Merck) was used. The column dimensions and the eluent(s) used are mentioned in each experiment separately.

Medium pressure liquid chromatography (MPLC)

A dosage pump FL1 with pulsation attenuator MPD 3 (both from Lewa company) was used. The detection was done using a UV/VIS Spectrometer 97.00 (Knauer company) and a differential refractometer connected to a plotter 41.21 (Knauer company). Type C columns filled with silica gel (column dimensions: 69 cm length 5 cm width, pressure 15-20 bar, flow 50-60 mL per minute, theoretical plate number 11500, paticle size 15-25 µm) were prepared according to G. Helmchen and B. Glatz.[143]

Solvents and reagents

All solvents and reagents used were purified and dried according to standard methods.[144]

7.2 Starting materials and suppliers

AllMgBr, ca. 2 M THF solution, Aldrich
LiHMDS, ca. 1 M THF solution, Fluka
AdBr, 99 % purity, Aldrich
Et$_2$AlCl, ca. 1 M *n*-hexane solution, Fluka
n-BuLi, ca. 15 % *n*-hexane solution, Merck

Benzyaldehyde oxime (**1**) was prepared as described in the literature.[57]

N-Benzylhydroxylamine (**2**) was obtained in accordance with the literature[58,59] in 72 % yield, m. p. 49-52°C (lit.:[58] 79 %, m. p. 58-59 °C).

1,2:5,6-Diisopropylidene-D-mannitol (**3**) was prepared according to the literature[27] in 49 % yield, m. p . 117-119 °C (lit. yield 54 %, m. p. 121.8-123.4 °C).

2,3-*O*-Isopropylidene-D-glyceraldehyde *N*-benzylnitrone (**4**) was synthesized according to a know procedure,[56] yield 68 %, m. p. 85-87 °C, $[\alpha]_D^{20}$ = 92 (c = 0.6, CHCl$_3$) [lit.: yield 86 %, m. p. 88 °C, $[\alpha]_D$ = 96.7 (c = 0.5, CHCl$_3$)].

Fluorenone *N*-(*p*-methoxyphenyl)-imine (**48**, 89 % yield) and benzophenone *N*-(*p*-methoxyphenyl)-imine (**49**, 69 % yield) were synthesized by Simone Vogt (Saring) and David Voelkel (lab assistants at the Institut für Organische Chemie, Universität Stuttgart).

The numbers of the experiments are given in order. The abbreviation and number given in the brackets correspond to the number in the lab journal, AB stands for Alevtina Baskakova.

7.3   Experiments concerning Chapter 2

(2S,3R)- and
(2S,3S)-N-Benzyl-3-(hydroxylamino)-1,2-O-isopropylidene-3-phenyl-
1,2-propanediol, (5a) and (5b)

5

a) Experiment 1 (AB 046). Addition of phenylmagnesium bromide to 2,3-*O*-isopropylidene-D-glyceraldehyde *N*-benzylnitrone (**4**) in the presence of ZnBr$_2$[51,53,55]

Typical procedure for the addition of Grignard reagents to the N-benzylnitrone 4 in the presence of a Lewis acid (TP 1)

To a well-stirred solution of the *N*-benzylnitrone **4** (0.47 g, 2.0 mmol) in absolute ether (40 mL) anhydrous ZnBr$_2$ (0.45 g, 2.0 mmol) was added in one portion at room temperature. The resulting mixture was stirred for 10 min under nitrogen, then it was cooled to –60 °C, treated with phenylmagnesium bromide (3 mL of ca. 1 M THF solution, ca. 3 mmol), and stirred for 6 h at this temperature. The reaction was quenched by addition of 1 N aqueous NaOH solution (13 mL) and allowed to stir an additional 20 min. The aqueous layer was separated and extracted with ether (3 × 15 mL). The combined ethereal layers were washed with brine (2 × 20 mL), dried over Na$_2$SO$_4$, and concentrated *in vacuo* (19 mbar) to give 0.56 g (89 %) of yellow oil. This was purified by flash chromatography on silica gel (30 g, 9 cm × 3 cm column), elution with ethyl acetate/petroleum ether mixture (50:50, v/v). From this, 0.30 g (48 %; lit.:[51,53] 86 %) of a spectroscopically pure yellow oil was obtained, which solidified on standing.

According to the ^{13}C NMR spectrum, the *threo/erythro* (**5a**/**5b**) ratio was 58:42 (Lit.:[51] 78:22; lit.:[53] 90:10); according to the HPLC analysis the ratio was 57:43.

Major isomer **5a** (*threo*):
^{13}C NMR (75.1 MHz, CDCl$_3$): δ = 25.8 and 26.8[2 q, C(CH$_3$)$_2$], 61.6 (t, C-1), 67.5 (d, C-3), 73.1 (d, CH$_2$C$_6$H$_5$), 76.5 (d, C-2), 110.1 [s, C(CH$_3$)$_2$], 127.2, 128.3, 128.5, 129.5, 130.1 (5 d, o-, m-, p-C of 2 C$_6$H$_5$), 136.3 and 137.8 (2 s, i-C of 2 C$_6$H$_5$).

Minor isomer **5b** (*erythro*):
^{13}C NMR (75.1 MHz, CDCl$_3$): δ = 25.6 and 26.5 [2 q, C(CH$_3$)$_2$], 62.1 (t, C-1), 68.4 (d, C-3), 73.7 (t, CH$_2$C$_6$H$_5$), 76.4 (d, C-2), 109.1 [s, C(CH$_3$)$_2$], 127.2, 128.1, 128.2, 128.3, 129.4, 130.2 (6 d, o-, m-, p-C of 2 C$_6$H$_5$, overlapping with those of the major isomer), 136.4 and 137.9 (2 s, i-C of 2 C$_6$H$_5$).

MPLC separation of this mixture was not successful [elution with ethyl acetate/petroleum ether (25:75, v/v), detection with an UV lamp (λ = 260 nm)].

b) Experiment 2 (AB 047). Addition of phenylmagnesium bromide to 2,3-*O*-isopropylidene-D-glyceraldehyde *N*-benzylnitrone in THF[51,53]

Typical procedure for the addition of organometallic reagents without additives to the N-benzylnitrone 4 (TP 2)[51,53]

A well-stirred solution of the *N*-benzylnitrone **4** (0.40 g, 1.7 mmol) in absolute THF (40 mL) was cooled to –60 °C, treated with PhMgBr (ca. 3.4 mmol, 3.4 mL of 1 M THF solution, Aldrich), and stirred for 6 h at this temperature. The reaction was quenched by addition of 1 N aqueous NaOH solution (13 mL) and allowed to stir for 20 min. The aqueous layer was separated and extracted with ether (3 × 15 mL). The combined ethereal layers were washed with brine (2 × 20 mL), dried over Na$_2$SO$_4$, and concentrated *in vacuo* to give 0.56 g (89 %) of a yellow oil. This was purified by flash chromatography on silica gel (30 g), elution with ethyl acetate/petroleum ether (1:2, v/v), yielding 437 mg (82 %, lit.:[51,53] 84 %) of the *threo/erythro* (**5a**/**5b**) mixture in form of a yellow, spectroscopically pure oil. According to the ^{13}C NMR spectra, the *threo/erythro* ratio was 75:25 (Lit.:[51] 73:27; lit.:[53] 80:20).

MPLC separation of this mixture was possible by elution with isopropanol/petroleum ether (3:97, v/v), detection with an UV lamp (λ = 254 nm).

(2S,3R)-N-Benzyl-3-(hydroxylamino)-1,2-O-isopropylidene-3-phenyl-1,2-propanediol
(5a) (the *threo*-isomer) was isolated in 33 % yield (175 mg) as a colourless, spectroscopically
and analytically pure oil.

$[\alpha]_D^{20}$ = –10.8 (c = 1.00, CHCl$_3$); lit.:[51,53,55] $[\alpha]_D^{20}$ = –6.5 (c = 1.00, CHCl$_3$).

C$_{19}$H$_{23}$NO$_3$	calcd	C 72.82	H 7.40	N 4.47
(313.4)	found	C 72.80	H 7.32	N 4.28

^{1}H NMR (300.1 MHz, CDCl$_3$): δ = 1.38 and 1.40 [2 s, 3 H each, C(CH$_3$)$_2$], 3.43 ("dd",
$^2J_{1a,1b}$ = 8.7, $^3J_{1a,2}$ = 6.9 Hz, 1 H, 1-H$_a$), 3.65 ("dd", $^2J_{1a,1b}$ = 8.6, $^3J_{1b,2}$ = 6.3 Hz, 1 H, 1-H$_b$), 3.66
and 3.81 (A, B from AB, $^2J_{A,B}$ = 13.2 Hz, 2 H, CH$_A$H$_B$C$_6$H$_5$), 3.73 ("d", $^3J_{2,3}$ = 8.7 Hz, 1 H, 3-H),
4.77 ("ddd", $^3J_{1a,2}$ = 6.9, $^3J_{1b,2}$ = 6.3, $^3J_{2,3}$ = 8.7 Hz, 1 H, 2-H), 6.22 (s, 1 H, NOH), 7.20–7.32
(m, 10 H, o-, m-, p-H of 2 C$_6$H$_5$).

^{13}C NMR (75.5 MHz, CDCl$_3$): δ = 25.9 and 26.8 [2 q, C(CH$_3$)$_2$], 61.7 (t, C-1), 67.5 (d, C-3),
72.9 (d, CH$_2$C$_6$H$_5$), 77.0 (d, C-2), 110.0 [s, C(CH$_3$)$_2$], 127.2, 128.3, 128.5, 129.5, 129.8 (5 d,
o-, m-, p-C of 2 C$_6$H$_5$), 136.2 and 137.8 (2 s, i-C of 2 C$_6$H$_5$).

The ^{1}H and ^{13}C NMR data were in agreement with the literature data.[53]

(2S,3S)-N-Benzyl-3-(hydroxylamino)-1,2-O-isopropylidene-3-phenyl-1,2-propanediol
(5b) (the *erythro*-isomer) was isolated in a yield of 67 mg (13 %) as a colourless,
spectroscopically, and analytically pure solid (m. p. 122–124 °C; lit.:[53] 141–143 °C).

$[\alpha]_D^{20}$ = –20 (c = 1.0, CHCl$_3$); lit.:[51,53,55] $[\alpha]_D^{20}$ = –17.5 (c = 1.0, CHCl$_3$)

C$_{19}$H$_{23}$NO$_3$	calcd	C 72.82	H 7.40	N 4.47
(313.4)	found	C 72.53	H 7.40	N 4.49

IR (neat): ν = 3382 (s, b, OH), 2989 (w), 2882 (w), 1493 (m), 1372 (m), 1257 (m), 1220 (s), 1156 (s, C–O), 1077 (s), 1022 (s), 1010 (s), 866 (s), 698 (vs) cm^{-1}.

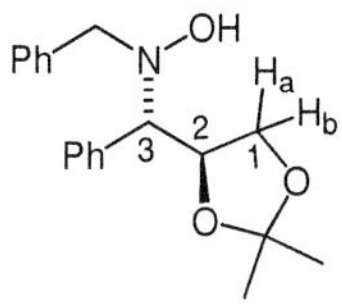

^{1}H NMR (300.1 MHz, CDCl$_3$): δ = 1.24 and 1.29 [2 s, 3 H each, C(CH$_3$)$_3$], 3.52 and 3.67 (A, B from AB, ^2J = 13.4 Hz, 2 H, CH$_A$H$_B$C$_6$H$_5$), 3.70 ("d", $^3J_{2,3}$ = 7.1 Hz, 1 H, 3-H), 3.94 ("dd", $^2J_{1a,1b}$ = 8.4, $^3J_{1a,2}$ = 6.5 Hz, 1 H, 1-H$_a$), 4.15 ("dd", $^2J_{1a,1b}$ = 8.6, $^3J_{1b,2}$ = 6.2 Hz, 1 H, 1-H$_b$), 4.75 ("q", $^3J_{1a,2}$ = $^3J_{1b,2}$ = $^3J_{2,3}$ = 6.9 Hz, 1 H, 2-H), 4.93 (bs, 1 H, NOH), 7.21–7.34 (m, 10 H, o-, m-, p-H of 2 C$_6$H$_5$).

^{13}C NMR (75.5 MHz, CDCl$_3$): δ = 25.5 and 26.5 [2 q, C(CH$_3$)$_2$], 62.2 (t, C-1), 68.3 (d, C-3), 73.7 (t, CH$_2$C$_6$H$_5$), 76.5 (d, C-2), 109.6 [s, C(CH$_3$)$_2$], 127.4 and 128.1 (2 d, p-C of 2 C$_6$H$_5$), 128.2, 128.3, 129.4, 130.2 (4 d, o-, m-C of 2 C$_6$H$_5$), 136.4 and 137.8 (2 s, i-C of 2 C$_6$H$_5$).

c) Experiment 3 (AB 48). Addition of phenyllithium to 2,3-O-isopropylidene-D-glyceraldehyde N-benzylnitrone (**4**)[54]

The reaction was performed according to TP 2:

N-Benzylnitrone **4** 0.40 g, 1.7 mmol
Abs. ether 30 mL
PhLi ca. 3.2 mmol, 1.7 mL of ca. 1.9 M dibutyl ether solution
Addition at –85 °C dropwise over 15 min
Stirring at -80 °C for 40 min
Quenching by saturated aqueous NH$_4$Cl solution (15 mL)

The crude product was purified by flash chromatography on silica gel (30 g, 9 cm × 3 cm column), elution with ethyl acetate/petroleum ether mixture (40:60, v/v) to yield 390 mg (73 %) of a mixture of isomers as a yellow, spectroscopically pure oil. Diastereomeric ratio **5a:5b** was 85:15 (according to ^{13}C NMR spectra).

The MPLC separation of this mixture was done by elution with isopropanol/petroleum ether 3:97 (v/v), detection with an UV lamp (λ = 255 nm).

(2*S*,3*R*)-*N*-Benzyl-3-(hydroxylamino)-1,2-*O*-isopropylidene-3-phenyl-1,2-propanediol
(5a) (the *threo*-isomer) was isolated as a colourless, spectroscopically pure oil, yield 245 mg
(46 %).

$[\alpha]_D^{20} = -16$ (c = 1.37, CHCl$_3$); lit.:[51,53,55] $[\alpha]_D^{20} = -6.5$ (c = 1.00, CHCl$_3$)

For NMR data of this compound see above.

(2*S*,3*S*)-*N*-Benzyl-3-(hydroxylamino)-1,2-*O*-isopropylidene-3-phenyl-1,2-propanediol 5b
(the *erythro*-isomer) was isolated as a colourless, spectroscopically pure solid, yield 46 mg
(9 %)
$[\alpha]_D^{20} = -30$ (c = 2.03, CHCl$_3$); lit.:[51,53,55] $[\alpha]_D^{20} = -17.5$ (c = 1.00, CHCl$_3$).

For NMR data of this compound see above.

d) Experiment 4 (AB 50). Addition of phenylmagnesium bromide to 2,3-*O*-isopropylidene-D-
glyceraldehyde *N*-benzylnitrone (**4**) in the presence of ZnCl$_2$[51,53,55]

The experiment was performed according to TP 1:

N-Benzylnitrone **4** 0.40 g, 1.7 mmol

Abs. diethyl ether 40 mL

ZnCl$_2$ anhydrous 0.20 g, 1.7 mmol

PhMgBr ca. 2.6 mmol, 2.6 mL of 1 M THF solution

The yield of the crude product in form of a yellow oil was 65 mg (12 %). The diastereomeric
ratio **5a:5b** was 83:17 (^{13}C NMR).
The crude product has not been purified further.

e) Experiment 5 (AB 91). Addition of phenylmagnesium bromide to 2,3-*O*-isopropylidene-D-
glyceraldehyde *N*-benzylnitrone (**4**) in the presence of Et$_2$AlCl.

The experiment was carried out according to TP 1:

N-benzylnitrone **4** 2.0 g, 8.5 mmol

Et$_2$AlCl 8.5 mL of ca. 1 M hexane solution, ca. 8.5 mmol

Abs. ether 70 mL

PhMgBr ca. 12.8 mmol, 12.8 mL of 1 M THF solution

Stirring at –60 °C for 16 h

The work-up gave 2.43 g (91 %; lit.[51] 78 %) of the product **5b** as a spectroscopically pure yellow solid. The *threo/erythro* ratio was <5:95 (lit.:[51] 15:85) (according to the NMR data, see above). The crude product was purified by flash chromatography on silica gel (50 g, 3 cm column) eluting with ethyl acetate/petroleum ether 35:65 to give 1.87 g (70 %) of a colourless solid, m. p. 95–97 °C; lit.:[53,51] 141–143 °C.

The NMR data were in accordance with those from the Exp. 2.

(2S,3RS)-N-Benzyl-3-*tert*-butyl-3-hydroxylamino-1,2-O-isopropylidene-1,2-propanediol (6a/b)

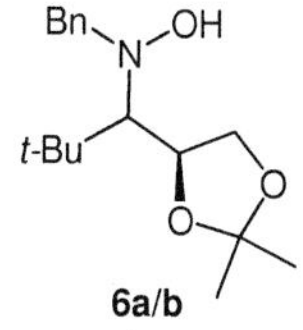

6a/b

a) Experiment 6 (AB 56). Addition of *tert*-butyllithium to 2,3-O-isopropylidene-D-glyceraldehyde N-benzylnitrone (**4**)[54]

The reaction was done according to TP 2:

N-benzylnitrone **4** 400 mg, 1.70 mmol
Abs. ether 35 mL
t-BuLi ca. 2.6 mmol, 1.5 mL of ca. 1.7 M pentane solution (Aldrich)
Addition at –85 °C within 15 min
Stirred at -80 °C for 40 min
Quenched by saturated aqueous NH$_4$Cl solution (15 mL)

The crude product was isolated as an orange oil (370 mg), which was purified by flash chromatography on silica gel (30 g, 9 cm × 3 cm column), elution with ethyl acetate/petroleum ether (35:65, v/v) to yield 100 mg (20 %) of a spectroscopically almost pure yellow oil. The ^{13}C NMR spectra showed a mixture of two diastereomers in ratio 75:25 (**6a:6b**).

The MPLC separation of this mixture did not succeed [elution with *i*-PrOH/petroleum ether (2:98, v/v), detection by UV (λ = 260 nm)].

b) Experiment 7 (AB 61). Addition of *tert*-butylmagnesium chloride to 2,3-O-isopropylidene-D-glyceraldehyde N-benzylnitrone (**4**)

The reaction was carried out according to TP 2:

N-Benzylnitrone **4** 0.50 g, 2.1 mmol

Abs. diethyl ether 30 mL

t-BuMgCl ca. 4.3 mmol, 2.2 mL of ca. 2 M ether solution

The crude product was isolated as a spectroscopically almost pure orange oil, 0.38 g ("62 %).

The ^{13}C NMR data of the crude product showed a diastereomeric ratio of 78:22 (**6a:6b**).

Main diastereomer (**6a**):

^{13}C NMR (75.5 MHz, CDCl$_3$): δ = 25.6 and 26.7 [2 q, C(CH_3)$_2$], 28.1 [q, C(CH_3)$_3$], 34.7 [s, C(CH$_3$)$_3$], 65.3 (t, $CH_2C_6H_5$), 70.0 (t, C-1), 74.0 (d, C-3), 75.9 (d, C 2), 108.0 [s, C(CH$_3$)$_2$], 127.0 (d, p-C of C$_6$H$_5$), 128.2 and 129.2 (2 d, o-, m-C of C$_6$H$_5$), 139.1 (s, i-C of C$_6$H$_5$).

Minor diastereomer (**6b**):

^{13}C NMR (75.5 MHz, CDCl$_3$): δ = 25.7 and 26.9 [2 q, C(CH_3)$_2$], 28.8 [q, C(CH_3)$_3$], 36.1 [s, C(CH$_3$)$_3$], 64.4 (t, $CH_2C_6H_5$), 68.5 (t, C-1), 74.2 (d, C-3), 75.6 (d, C 2), 108.0 [s, C(CH$_3$)$_2$, overlapping signal], 127.0 (d, p-C of C$_6$H$_5$), 128.2 and 129.2 (2 d, o-, m-C of C$_6$H$_5$), 139.1 (s, i-C of C$_6$H$_5$).

The crude product was purified by flash chromatography on silica gel (30 g, 9 cm × 3 cm column), elution with ethyl acetate/petroleum ether (35:65, v/v) to yield 0.33 g (53 %) of a mixture of two diastereomers as a yellow oil. This was separated by MPLC, elution with iPrOH/petroleum ether 3:97 (v/v), detection by UV (λ = 255 nm). Two fractions were obtained: a mixture of two diastereomers (45 mg, 7 %, diastereomeric ratio **6a:6b** 76:24) and the diastereomer **6a** as a spectroscopically and analytically pure colourless solid (70 mg, 11 %, m. p. 157-158 °C).

Analytical data of **6a**:

$[α]_D^{20}$ = 10.0 (c = 0.36, CHCl$_3$)

C$_{17}$H$_{27}$NO$_3$	calcd	C 69.59	H 9.28	N 4.77
(293.4)	found	C 69.93	H 8.65	N 4.73

^{1}H NMR (300.1 MHz, CDCl$_3$): δ = 0.95 [s, 9 H, C(CH_3)$_3$], 1.43 and 1.53 [2 s, 3 H each, C(CH$_3$)$_2$], 2.55 ("d", $^3J_{2,3}$ = 7.9 Hz, 1 H, 3-H), 3.72 ("t", $^2J_{1a,1b}$ = $^3J_{1a,2}$ = 8.3 Hz, 1 H, 1-H$_a$), 4.15 ("dd", $^2J_{1a,1a}$ = 8.1, $^3J_{1b,2}$ = 6.0 Hz, 1 H, 1-H$_b$), 4.15 and 4.39 (A, B of AB, $^2J_{A,B}$ = 13.4 Hz, 1 H each, CH_AH$_B$C$_6$H$_5$), 4.54 ("dt", $^3J_{1b,2}$ = 6, $^3J_{1a,2}$ = $^3J_{2,3}$ = 8.2 Hz, 1 H, 2-H), 5.30 (bs, 1 H, OH), 7.23–7.40 (m, 5 H, o-, m-, p-H of C$_6$H$_5$).

^{13}C NMR (75.5 MHz, CDCl$_3$): δ = 25.6 and 26.7 [2 q, C(CH$_3$)$_2$], 28.2 [q, C(CH$_3$)$_3$], 34.7 [s, C(CH$_3$)$_3$], 65.3 (t, CH$_2$C$_6$H$_5$), 69.9 (t, C-1), 74.0 (d, C-3), 75.9 (d, C 2), 108.0 [s, C(CH$_3$)$_2$], 127.0 (d, p-C of C$_6$H$_5$), 128.2 and 129.2 (2 d, o-, m-C of C$_6$H$_5$), 139.1 (s, i-C of C$_6$H$_5$).

(2S,3RS)-3-(1-Adamantyl)-3-(N-benzylhydroxylamino)-1,2-O-isopropylidene-1,2-propanediol (7b)

a) Experiment 8 (AB 081). Addition of 1-adamantylmagnesium bromide to 2,3-O-isopropylidene-D-glyceraldehyde N-benzylnitrone (**4**)

Typical procedure for the preparation of the ethereal solution of adamantylmagnesium bromide (TP 3)^{145}

A 50 ml two-necked flask was equipped with a condenser, connected to the vacuum line and charged with magnesium turnings (23.3 mmol, 0.56 g). The flask was preheated to 150 °C for 10 min under reduced pressure, then cooled under nitrogen, and absolute ether (4 mL) was added followed by addition of 3 drops of dibromoethane. To this suspension a solution of 1-bromoadamantane (4.66 mmol, 1.00 g) in absolute ether (6 mL) was carefully added within 20-30 min, keeping the temperature below the boiling point of ether. The suspension was stirred for 30 min at room temperature and heated under reflux for 1 h. After cooling to room temperature, the solution of 1-adamantylmagnesium bromide was used immediately in the next step.

The experiment was carried out according to TP 2:

N-Benzylnitrone **4** 0.55 g, 2.3 mmol

Abs. ether 40 mL

AdMgBr ca. 4.7 mmol

Stirring for 17 h at -60 °C (cryostate)

Quenched by sat. aqueous NH_4Cl solution (10 mL)

The crude product was isolated as a yellow solid, 0.92 g ("106 %"). This was purified by flash chromatography on silica gel (60 g, 18 cm × 3 cm column), elution with petroleum ether and then with ethyl acetate/petroleum ether (15:85, v/v) to yield 0.44 g (51 %) of a colourless solid, consisting of **7b:7a** mixture in 58:42 ratio (according to the ^{13}C NMR spectra).

Major diastereomer (**7b**, *erythro*):

^{13}C NMR (62.9 MHz, $CDCl_3$): δ = 25.7 and 27.00 [2 q, $C(CH_3)_2$], 28.8 (d, C-3', C-5', C-7'), 37.2 (t, C-4', C-6', C-9'), 38.1 (s, C-1'), 40.4 (t, C-2', C-8', C-10'), 64.7 (t, $CH_2C_6H_5$), 68.2 (t, C-1), 74.9 (d, C-3), 75.0 (d, C-2), 107.2 [s, $C(CH_3)_2$, 127.2 (d, *p*-C of C_6H_5), 128.3 and 129.0 (2 d, *o*-, *m*-C of C_6H_5), 139.4 (s, *i*-C of C_6H_5).

Minor diastereomer (**7a**, *threo*):

^{13}C NMR (62.9 MHz, $CDCl_3$): δ = 25.4 and 26.6 [2 q, $C(CH_3)_2$], 28.6 (d, C-3', C-5', C-7'), 36.7 (t, C-4', C-6', C-9'), 36.8 (t, C-2', C-8', C-10'), 39.9 (s, C-1'), 65.6 (t, $CH_2C_6H_5$), 70.1 (t, C-1), 75.0 (d, C-3), 108.0 [s, $C(CH_3)_2$], 127.1 (d, *p*-C of C_6H_5), 128.2 and 128.2 (2 d, *o*-, *m*-C of C_6H_5), 139.1 (s, *i*-C of C_6H_5).

The MPLC separation of this mixture was unsuccessful [elution with isopropanol/petroleum ether mixture (2:98, v/v), detection with an UV lamp, λ = 254 nm).

b) Experiment 9 (AB 084). Addition of adamantylmagnesium bromide to 2,3-*O*-isopropylidene-D-glyceraldehyde *N*-benzylnitrone (**4**) in the presence of Et_2AlCl[51,52,53]

Adamantylmagnesium bromide was prepared according to TP 3.

The reaction was done as described in the TP 1:

(*Z*)-*N*-Benzyl-2,3-*O*-isopropylidene-D-glyceraldehyde nitrone (**4**) 1.00 g, 4.3 mmol

Abs. diethyl ether 50 mL

Et_2AlCl ca. 4.3 mmol, 4.3 ml of ca. 1 M hexane solution

AdMgBr ca. 8.5 mmol

Stirring for 16 h at –60 °C (cryostat)

Quenched by 2.5 % aqueous KOH solution (20 mL)

The crude product was isolated as a yellow solid, 1.23 g ("78 %") which was purified by flash chromatography on silica gel (60 g, 18 cm × 3 cm column), elution with petroleum ether and then with ethyl acetate/petroleum ether (20:80, v/v) to yield 0.67 g ("42 %") of a spectroscopically pure colourless solid, m. p. 131–132 °C. According to the ^{13}C NMR spectra, the ratio **7b:7a** (*erythro:threo*) was >95:5.

$[\alpha]_D^{20} = 31$ (c = 0.67, CHCl₃)

C₂₃H₃₃NO₃	calcd	C 74.36	H 8.95	N 3.77
(371.5)	found	C 74.54	H 8.99	N 3.63

IR (neat): ν = 3417 (w, OH), 2899 (s), 2843 (s, NH₂), 1447 (m), 1113 (vs, C–O), 1008 (s), 909 (s), 695 (s) cm⁻¹.

^{1}H NMR (500.1 MHz, CDCl₃): δ = 1.39 and 1.48 [2 s, 3 H each, C(CH₃)₂], 1.68 (m, 6 H, 4'-H, 6'-H, 9-H), 1.78 (m, 6 H, 2'-H, 8'-H, 10'-H), 1.98 (m, 3 H, 3'-H, 5'-H, 7'-H), 2.58 ("d", $^3J_{2,3}$ = 5.6 Hz, 1 H, 3-H), 3.96 (A of AB, $^2J_{A,B}$ = 13.7 Hz, 1 H, CH$_A$H$_B$C₆H₅), 4.20 ("dd", $^2J_{1a,2}$ = 6.4, $^3J_{1b,2}$ = 8.1 Hz, 2 H, 1-H$_a$, 1-H$_b$), 4.24 (B of AB, ^2J = 13.7 Hz, 1 H, CH$_A$H$_B$C₆H₅), 4.39 (bs, 1 H, N-OH), 4.46 ("ddd", $^2J_{1a,2}$ = 6.4, $^3J_{1b,2}$ = 8.1, $^3J_{2,3}$ = 5.6 Hz, 1 H, 2-H), 7.24–7.40 (m, 5 H, o-, m-, p-H of C₆H₅).

^{13}C NMR (125.8 MHz, CDCl₃): δ = 25.7 and 26.9 [2 q, C(CH₃)₂], 28.7 (d, C-3', C-5', C-7'), 37.1 (t, C-4', C-6', C-9'), 40.4 (t, C-2', C-8', C-10'), 38.0 (s, C-1'), 64.6 (t, CH₂C₆H₅), 68.1 (t, C-1), 74.8 (d, C-3), 75.0 (d, C-2), 107.1 [s, C(CH₃)₂], 127.1 (d, p-C of C₆H₅), 128.3 and 128.9 (2 d, o-, m-C of C₆H₅), 139.3 (s, i-C of C₆H₅).

The assignment of the signals was done by means of C,H COSY experiments.

Experiment 10 (AB 85)

(2*S*,3*S*)-3-(*tert*-Butoxycarbonylamino)-1,2-di-*O*-isopropylidene-3-
phenyl-1,2-propanediol (8a)

BocHN

Ph

8a

Typical procedure for the hydrogenation of N-benzylhydroxylamines in the presence of Pd(OH)$_2$ (TP 4)[55]

Diastereomerically pure (2*S*,3*S*)-*N*-Benzyl-3-(hydroxylamino)-1,2-*O*-isopropylidene-3-phenyl-1,2-propanediol (**5b**, from AB 91) (1.80 g, 5.8 mmol) was dissolved in dry methanol (65 ml), then di-*tert*-butyl dicarbonate (2.51 g, 11.5 mmol) was added together with Et$_3$N (ca. 0.05 mL) and the resulting solution was transferred into a hydrogenation vessel. This was flushed with nitrogen and Pd(OH)$_2$/C 10 % (190 mg) was added. The reaction was carried out under hydrogen pressure (4.8 bar) for 3 days. Then the suspension was filtered through a pad of celite/silica gel (ca. 2 cm, celite on top) and the adsorbent was washed with ethyl acetate (2 × 20 mL). The filtrate was concentrated *in vacuo* (18 mbar) to give 1.64 g (93 %) of a colourless solid, which was purified by column chromatography on silica gel (75 g, 20 cm × 3.5 cm column), elution with ethyl acetate/petroleum ether (15:85, v/v) and 2 % of Et$_3$N. From this 0.99 g (56 %) of **8a** in form of a spectroscopically and analytically pure, colourless solid was obtained (m. p. 93–102 °C; lit.:[55] 115 °C).

$[\alpha]_D^{20} = 23.0$ (c = 0.34, CHCl$_3$); lit.:[55] $[\alpha]_D^{20} = 24.2$ (c = 1.0, CHCl$_3$)

C$_{17}$H$_{25}$NO$_4$	calcd.	C	66.43	H	8.20	N	4.56
(307.4)	found	C	66.41	H	8.20	N	4.37

IR (neat): ν = 3372 (m, NH), 2976 (m), 1692 (vs, C=O), 1519 (vs), 1497 (vs), 1389 (m), 1366 (s), 1350 (s), 1311 (m), 1246 (vs), 1162 (vs), 1057 (vs), 1041 (s), 918 (m), 892 (s), 859 (s), 770 (s), 761 (s), 700 (s), 631 (m) cm^{-1}.

^{1}H NMR (300.1 MHz, CDCl$_3$): δ (ppm) = 1.33 and 1.34 [2 s, 3 H each, C(CH$_3$)$_2$], 1.41 [s, 9 H, C(CH$_3$)$_3$], 3.73 ("dd", $^2J_{1a,1b}$ = 8.7, $^3J_{1a,2}$ = 6.1 Hz,1 H, 1-H$_a$), 3.95 ("dd", $^2J_{1a,1b}$ = 8.7, $^3J_{1b,2}$ = 6.5

Hz, 1 H, 1-H$_b$), 4.39 ("q", $^3J_{1a,2}$ = $^3J_{1b,2}$ = $^3J_{2,3}$ = 5.8 Hz,1 H, 2-H), 4.78 (bs, 1 H, N-H), 5.18 (bs, 1 H, 3-H), 7.21–7.31 (m, 5 H, o-, m-, p-H of C$_6$H$_5$).

^{13}C NMR (75.5 MHz, CDCl$_3$): δ (ppm) = 25.2 and 26.3 [2 q, C(CH_3)$_2$], 28.3 [q, C(CH_3)$_3$], 56.5 (d, C-3), 65.9 (t, C-1), 77.9 (d, C-2), 79.8 [C(CH$_3$)$_3$], 109.8 [C(CH$_3$)$_2$], 127.5 (d, p-C of C$_6$H$_5$), 127.6 and 128.4 (2 d, o-, m-C of C$_6$H$_5$), 138.7 (s, i-C of C$_6$H$_5$), 155.4 (C=O).

The NMR data were in agreement with the literature data.[55]

Experiment 11 (AB 089)

(2S,3R)-3-($tert$-Butoxycarbonylamino)-1,2-di-O-isopropylidene-3-phenyl-1,2-propanediol (8b)

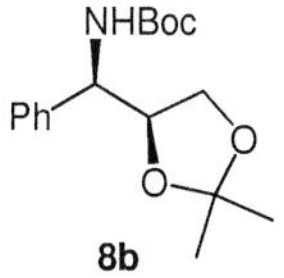

8b

The hydrogenation was carried out according to TP 4:

(2S,3R)-3-(Benzylhydroxylamino)-1,2-O-isopropylidene-3-phenyl-1,2-propanediol   **(5b)**
720 mg, 2.30 mmol
Methanol 35 mL
Di-$tert$-butyl dicarbonate 750 mg, 3.40 mmol
Pd(OH)$_2$/C 10 % 76 mg

The crude product was obtained as a colourless solid, which was purified by chromatography on silica gel (40 g, column 13 cm × 3 cm), elution with ethyl acetate/petroleum ether/Et$_3$N (15:85:2, v/v). From this, 480 mg (68 %) of **8b** in the form of a spectroscopically and analytically pure, colourless solid was obtained (m. p. 90–95 °C, lit.:[55] sticky oil).

$[α]_D^{20}$ = –27.5 (c = 0.26, CHCl$_3$); lit.:[55] $[α]_D^{20}$ = –27.5 (c = 0.67, CHCl$_3$)

C$_{17}$H$_{25}$NO$_4$	calcd.	C	66.43	H	8.20	N	4.56
(307.4)	found	C	66.55	H	8.17	N	4.45

IR (neat): ν = 3336 (w, NH), 2984 (m), 1706 (vs, C=O), 1521 (s), 1497 (vs), 1365 (m), 1242 (s), 1159 (vs), 1065 (s), 1045 (s), 1005 (s), 860 (s), 833 (m), 701 (vs), 621 (m) cm^{-1}.

^{1}H NMR (300.1 MHz, CDCl$_3$): δ = 1.32 and 1.46 [2 s, 3 H each, C(CH$_3$)$_2$], 1.39 [s, 9 H, C(CH$_3$)$_3$], 3.76 ("dd", $^2J_{1a,1b}$ = 8.4, $^3J_{1a,2}$ = 6.7 Hz,1 H, 1-H$_a$), 3.84 ("dd", $^2J_{1a,1b}$ = 8.4, $^3J_{1b,2}$ = 6.6 Hz, 1 H, 1-H$_b$), 4.32 ("dd", $^3J_{1a,2}$ = $^3J_{1b,2}$ = 6.6, $^3J_{2,3}$ = 10.7 Hz,1 H, 2-H), 4.68 (bs, 1 H, N-H), 5.38 ("d", $^3J_{2,3}$ = 8.2 Hz, 1 H, 3-H), 7.21–7.31 (m, 5 H, o-, m-, p-H of C$_6$H$_5$).

^{13}C NMR (75.5 MHz, CDCl$_3$): δ = 25.2 and 26.5 [2 q, C(CH$_3$)$_2$], 28.3 [q, C(CH$_3$)$_3$], 56.7 (d, C-3), 66.9 (t, C-1), 78.2 (d, C-2), 79.6 [s, C(CH$_3$)$_3$], 109.8 [s, C(CH$_3$)$_2$], 127.0 (d, p-C of C$_6$H$_5$), 127.5, and 128.5 (2 d, o-, m-, p-C of C$_6$H$_5$), 140.6 (s, i-C of C$_6$H$_5$), 155.6 (C=O).

The NMR data were in agreement with the literature data.[55]

Experiment 12 (AB 97)

(2S,3R)-3-(tert-Butoxycarbonylamino)-3-phenyl-1,2-propanediol (9)

According to a literature procedure,[47] to (2S,3S)-3-(tert-butoxycarbonylamino)-1,2-O-isopropylidene-3-phenyl-1,2-propanediol (**8a**) (600 mg, 1.97 mmol) in methanol (15 mL) p-toluenesulfonic acid was added (25 mg, 0.15 mmol) and the resulting solution was stirred for 17 h at ambient temperature. Then the solvent was removed in vacuo (20 mbar), the residual oil was treated with saturated aqueous NaHCO$_3$ solution, and the resulting suspension was extracted with dichloromethane (4 × 10 mL). The combined organic layers were dried over Na$_2$SO$_4$ and concentrated in vacuo to give 590 g ("107 %") of a colourless oil, which was purified by chromatography on silica gel (25 g, 12 cm × 2 cm column), elution with methanol/dichloromethane 95:5 (v/v). The diol **9** was isolated with a yield of 310 mg (59 %) as a colourless, analytically pure solid (m. p. 108–110 °C, lit.:[53] 116–118 °C).

$[\alpha]_D^{20}$ = 51 (c = 0.98, CHCl$_3$); lit.:[53] $[\alpha]_D$ = 51 (c = 0.65, CHCl$_3$)

| $C_{14}H_{21}NO_4$ | calcd | C 62.90 | H 7.92 | N 5.24 |
| (267.3) | found | C 62.84 | H 7.98 | N 5.16 |

IR (neat): ν = 3376 (m, NH), 1681 (vs, C=O), 1518 (vs), 1391 (m), 1366 (m), 1343 (m), 1288 (s), 1249 (s), 1166 (vs), 1064 (m), 1033 (s), 1008 (vs), 887 (s), 758 (s), 703 (vs), 629 (s), 590 (vs), 575 (s) cm^{-1}.

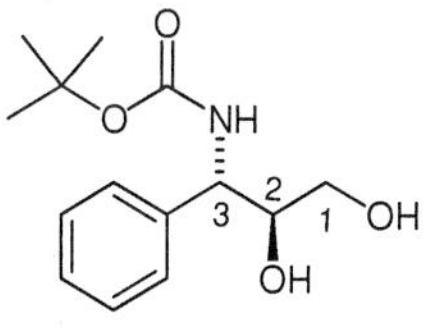

^{1}H NMR (300.1 MHz, CDCl$_3$): δ = 1.42 [s, 9 H, C(CH$_3$)$_3$], 2.96 and 3.34 (2 bs, 1 H each, 2 OH), 3.68 ("d", $^3J_{1,2}$ = 3.0 Hz, 2 H, 1-H), 3.83 ("dd", $^3J_{1,2}$ = 3.2, $^3J_{2,3}$ = 7.5 Hz, 1 H, 2-H), 4.69 ("t", $^3J_{1,NH}$ = $^3J_{2,3}$ = 7.8 Hz, 1 H, 3-H), 5.40 (bs, 1 H, NH), 7.27–7.38 (m, 5 H, o-, m-, p-H of C$_6$H$_5$).

The recorded ^{1}H NMR spectrum showed slight deviations with those from the literature:[53]

^{1}H NMR (300.1 MHz, CDCl$_3$): δ = 1.40 (s, 9 H), 2.40 (2 bs, 2 H), 3.52–3.60 (m, 2 H), 3.75–3.79 (m, 1 H), 4.54 (t, ^3J = 7.7 Hz, 1 H), 5.14 (bs, 1 H), 7.29–7.41 (m, 5 H).[53]

^{13}C NMR (75.5 MHz, CDCl$_3$): δ = 28.3 [2 q, C(CH$_3$)$_3$], 56.8 (d, C-3), 63.2 (t, C-1), 74.1 (d, C-2), 80.4 [s, C(CH$_3$)$_3$], 127.9 (d, p-C of C$_6$H$_5$), 127.6, and 128.8 (2 d, o-, m-C of C$_6$H$_5$), 139.0 (s, i-C of C$_6$H$_5$), 156.3 (s, C=O).

The ^{13}C NMR spectrum was in a good accordance with the literature data.[53]

Experiment 13 (AB 139)

(S)-N-(tert-Butoxycarbonyl)-phenylglycine (10)

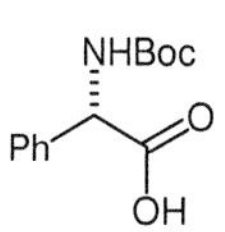

10

Oxidative transformation of a diol moiety into a carboxy group (TP 5)[61,62] *tert*-Butyl (2*S*,3*S*)-2,3-dihydroxy-1-phenylpropylcarbamate (**9**) (440 mg, 1.65 mmol) was dissolved in methanol/water (1:1, 8 mL), the solution was cooled to 0 °C, and NaIO$_4$ (424 mg, 1.98 mmol) was added under vigorous stirring. The suspension was left with stirring at room temperature for 30 min, diluted with water (8 mL) and extracted with diethyl ether (3 × 10 mL). The combined organic layers were washed with brine, dried over Na$_2$SO$_4$, and concentrated *in vacuo* (200 mbar, cold water bath). The residual colourless oil was dissolved in a mixture of *tert*-butanol (5 mL) and 2-methyl-2-butene (3.5 mL). To this, an aqueous solution of NaClO$_2$ (300 mg, 3.30 mmol) and KH$_2$PO$_4$ (450 mg, 3.30 mmol) was added dropwise within 30 min at 0 °C. The resulting suspension was stirred overnight (16 h), followed by addition of a solution NaClO$_2$ (150 mg, 1.65 mmol) and KH$_2$PO$_4$ (225 mg, 1.65 mmol) in water, and the suspension was stirred for an additional 2 h. Then 20 % aqueous NaOH solution was added to reach pH>9, and the organic solvents were removed *in vacuo* (100 mbar). The residue was diluted with water (5 mL), washed with diethyl ether (2 × 10 mL), and acidified by addition of saturated aqueous KHSO$_4$ solution to reach pH 1-2 (with cooling in an ice-bath). The solid formed was extracted with ethyl acetate (4 × 10 mL). The combined organic layers were washed with saturated aqueous Na$_2$SO$_3$, dried over Na$_2$SO$_4$, and concentrated to give 334 mg (81 %) of a yellowish oil. This was purified by column chromatography on silica gel (40 g, 9 cm × 4 cm), elution with methanol/dichloromethane 8:92 (v/v). The yield of the product as a spectroscopically and analytically pure, colourless solid was 320 mg (53 %), m. p. 42–45 °C (lit.:[64a] 88–91 °C).

$[\alpha]_D^{20}$ = + 112.4 (c = 0.88, EtOH); lit.:[64a] $[\alpha]_D^{20}$ = + 144 (c = 1, EtOH); lit.:[64b] $[\alpha]_D^{24}$ = + 135 (c = 0.8, CHCl$_3$)

C$_{13}$H$_{17}$NO$_4$	calcd	C 62.14	H 6.82	N 5.57
(251.3)	found	C 62.24	H 6.99	N 5.17

IR (neat): ν = 3363 (m), 2925 (vs), 2854 (vs), 1715 (vs), 1688 (vs), 1523 (s), 1456 (s), 1169 (m) cm^{-1}.

^{1}H NMR (300.1 MHz, CDCl$_3$): δ = 1.21 [s, 9 H, C(CH$_3$)$_3$], 5.11 (d, $^3J_{2,NH}$ = 5.2 Hz, 1 H, 2-H), 7.28–7.47 (m, 5 H, o-, m-, p-H of C$_6$H$_5$), 8.06 (d, $^3J_{2,NH}$ = 5.1 Hz, 1 H, NH), 10.90 (bs, 1 H, COOH).

^{13}C NMR (75.5 MHz, CDCl$_3$): δ = 28.0 [q, C(CH$_3$)$_3$], 58.9 (d, C-2), 81.7 [s, C(CH$_3$)$_3$], 127.2 (d, p-C of C$_6$H$_5$), 128.0 and 128.5 (2 d, o-, m-C of C$_6$H$_5$), 138.3 (s i-C of C$_6$H$_5$), 157.0 (s, C=O of Boc), 173.5 [s, C(O)OH].

Experiment 14 (AB 142)

(S)-Phenylglycine hydrochloride (11·HCl)

According to the literature,[146] (S)-N-(tert-butoxycarbonyl)-phenylglycine (**10**) (0.31 g, 1.24 mmol) was stirred in trifluoroacetic acid (3 ml) overnight (20 h). The acid was removed in vacuo (20 mbar), the residual solid was dissolved in water (ca. 10 mL) and put onto an ion-exchange column (Dowex 50WX8 resin, H$^+$ - form, 5 cm × 1 cm).

The resin was activated as follows:[103]

The ion resin in a column was washed successively with 50 mL of distilled water, 1 N ammonia, water (till neutralisation), 1 N hydrochloric acid, water (till neutralisation), methanol and water.

The column was washed successively with 50 mL of water, methanol and water and then the free amino acid was collected by elution with 1 N aqueous ammonia (100 mL). The yield of the (S)-phenylglycine (**11**) as a colourless solid was 0.11 g (56 %).

[α]$_D^{20}$ = + 137 (c = 0.71, 1 N HCl); lit.:[146] [α]$_D$ = + 124 (c = 0.5, 1 N HCl)

(S)-Phenylglycine (68 mg, 0.45 mmol) was dissolved in concentrated hydrochloric acid (3 ml), and the solution was stirred overnight (20 h) at room temperature. Removal of the acid (20 mbar) gave the product **11·HCl** as a spectroscopically pure and analytically almost pure, colourless solid (83 mg, quant.).

C$_8$H$_{10}$ClNO$_2$	found	C 49.06	H 5.44	N 7.41	Cl 19.28
(187.6)	calcd	C 51.21	H 5.37	N 7.47	Cl 18.90

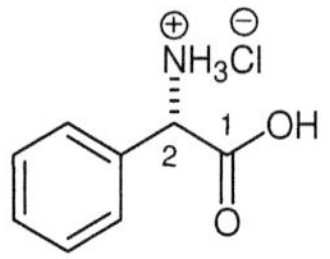

^{1}H NMR (300.1 MHz, D$_2$O): δ = 5.12 (s, 2 H, 2-H), 7.44–7.50 (m, 5 H, *o*-, *m*-, *p*-H of C$_6$H$_5$).

^{13}C NMR (75.5 MHz, D$_2$O): δ = 56.7 (d, C-2), 128.0 (d, *p*-C of C$_6$H$_5$), 129.6 and 130.2 (2 d, *o*-, *m*-C of C$_6$H$_5$), 131.6 (s, *i*-C of C$_6$H$_5$), 171.0 (C=O).

Experiment 15 (AB 158) cf. lit.[65,66]

(*S*)-2-*tert*-Butoxycarbonylamino-1-phenylethanol (12)

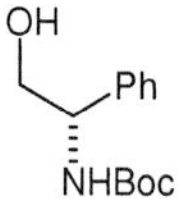

12

The diol **9** (0.500 g, 1.85 mmol) was dissolved in methanol/water (50:50, 10 mL), and sodium metaperiodate (0.486 g, 2.22 mmol) was added at 0 °C. The reaction was complete within 30 min, and the mixture was diluted with water (5 mL) and extracted with diethyl ether (3 × 15 mL). The organic phases were dried and concentrated *in vacuo* (150 mbar, cold water bath) to give a colourless oil which was dissolved in methanol (7 mL). Then sodium borohydride (0.176 g, 4.63 mmol) was added in small portions at 0 °C. Stirring was continued for 2 h at ambient temperature and the reaction was quenched by saturated aqueous NaHCO$_3$ solution (7 mL). The aqueous layer was separated and extracted with dichloromethane (3 × 20 mL). Drying (Na$_2$SO$_4$) and evaporation of the solvent gave 0.387 g (88 %) of a colourless solid, which was recrystallized from diethyl ether/petroleum ether to give 0.333 g (75 %) of the spectroscopically and analytically pure **12**, m. p. 116–131 °C.

[α]$_D^{20}$ = 36 (c = 0.96, CHCl$_3$); lit.:[66] [α]$_D^{20}$ = -37.0 (c = 1, CHCl$_3$) for the corresponding (*R*)-enantiomer.

C$_{13}$H$_{19}$NO$_3$	calcd	C 65.80	H 8.07	N 5.90
(237.3)	found	C 65.90	H 8.07	N 5.88

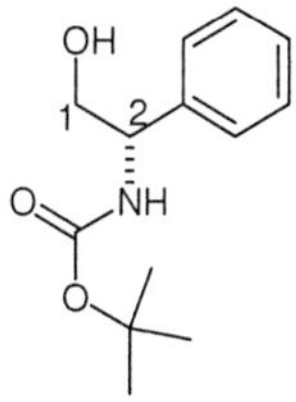

^{1}H NMR (300.1 MHz, CDCl$_3$): δ = 1.42 [s, 9 H, C(CH$_3$)$_3$], 2.16 (bs, 1 H), 3.48 (bs, 2 H), 4.75 (bs, 1 H), 5.28 (s, 1 H, NH), 7.24–7.37 (m, 5 H, o-, m-, p-H of C$_6$H$_5$).

^{13}C NMR (75.5 MHz, CDCl$_3$): δ = 28.3 [q, C(CH$_3$)$_3$], 56.9 (d, C-2), 66.7 (t, C-1), 80.8 [s, C(CH$_3$)$_3$], 127.7 (d, p-C of C$_6$H$_5$), 126.6 and 128.7 (2 d, o-, m-C of C$_6$H$_5$), 139.6 (i-C of C$_6$H$_5$), 156.2 (s, C=O).

Experiment 16 (AB 87)

(2S,3S)-3-(1-Adamantyl)-3-(tert-butoxycarbonylamino)-1,2-O-isopropylidene-1,2-propanediol (13)

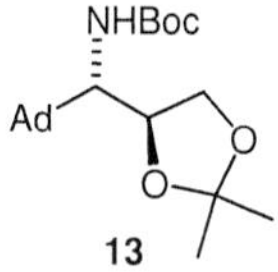

The hydrogenation was performed as described in TP 4:

N-Benzylhydroxylamine **7b** 320 mg, 0.86 mmol
Methanol 10 mL
Di-tert-butyl dicarbonate 282 mg, 1.29 mmol
Pd(OH)$_2$/C 40 mg
Reaction time 3 d

The crude product, a colourless semi-solid, was chromatographed on silica gel (10 g, 1 cm × 4 cm), elution with ethyl acetate/petroleum ether/triethylamine (50:50:2, v/v/v). From this, 192 mg (61 %) of the amine product **13** were isolated as an analytically pure, colourless solid, m. p. 109–114 °C.

$[\alpha]_D^{20}$ = 20 (c = 1.54, CH$_3$OH)

C$_{21}$H$_{35}$NO$_4$	calcd	C 69.01	H 9.65	N 3.83
(365.5)	found	C 68.77	H 9.58	N 3.57

IR (neat): ν = 2902 (vs, b), 2849 (vs), 1678 (vs, amide I), 1525 (s, amide II), 1452 (m), 1365 (s), 1242 (s), 1161 (vs), 1049 (vs, b, C-O), 1011 (vs, C–O), 940 (s), 827 (s) cm^{-1}.

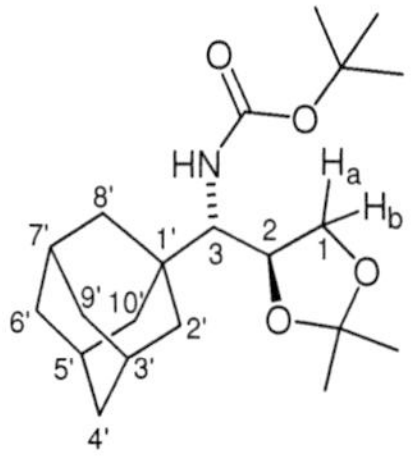

^{1}H NMR (500.2 MHz, CDCl$_3$): δ = 1.34 and 1.38 [2 s, 3 H each, C(CH$_3$)$_2$], 1.44 [s, 9 H, C(CH$_3$)$_3$], 1.51–1.71 (m, 12 H, 2'-H, 4'-H, 6'-H, 8'-H, 9'-H, 10'-H), 1.98 (bs, 3 H, 3'-H, 5'-H, 7'-H), 3.55 ("dd", $^{3}J_{2,3}$ = 6.2, $^{3}J_{2,NH}$ = 10.8 Hz, 1 H, 3-H), 3.67 ("dd", $^{2}J_{1a,1b}$ = 8.4, $^{3}J_{1a,2}$ = 7.7 Hz, 1 H, 1-H$_a$), 3.98 ("dd", $^{2}J_{1a,1b}$ = 8.4, $^{3}J_{1b,2}$ = 6.4 Hz, 1 H, 1-H$_b$), 4.23 ("dt", $^{3}J_{1a,2}$ = 7.5, $^{3}J_{1b,2}$ = $^{3}J_{2,3}$ = 6.3 Hz, 1 H, 2-H), 4.45 (d, $^{3}J_{3,NH}$ = 10.8 Hz, 1 H, NH).

^{13}C NMR (125.8 MHz, CDCl$_3$): δ = 25.7 and 26.4 [2 q, C(CH$_3$)$_2$], 28.3 [q, C(CH$_3$)$_3$], 28.4 (d, C-3', C-5', C-7'), 35.9 (s, C-1'), 36.9 (t, C-4', C-6', C-9'), 39.1 (t, C-2', C-8', C-10'), 58.1 (d, C-3), 66.8 (t, C-1), 74.7 (d, C-2), 79.2 [s, C(CH$_3$)$_3$], 108.8 [s, C(CH$_3$)$_2$], 156.2 (C=O).

The assignment of signals was done by means of C,H COSY experiments.

Experiment 17 (AB 159)

(2S,3S)-3-(1-Adamantyl)-3-[N-(Z)-benzylidene-N-oxyamino]-1,2-isopropylidenepropane-1,2-diol (14)

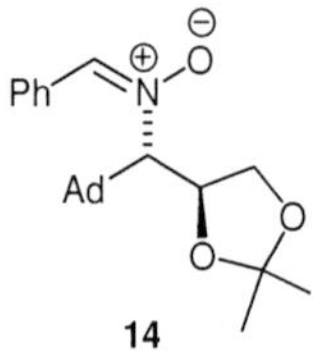

14

The hydrogenation was performed according to TP 4:

N-Benzylhydroxylamine **7b** 2.10 g, 5.1 mmol

Methanol/THF mixture 50 mL:15 mL

Di-tert-butyl dicarbonate 1.67 g, 7.65 mmol

Pd(OH)$_2$/C 100 mg

The reaction was run for ca. 68 h. The crude product, a colourless semi-solid, was chromatographed on silica gel, elution with ethyl acetate/petroleum ether/triethylamine (50:50:2, v/v/v). From this, 0.66 g ("63 %") of the starting material **7b** in a mixture with **14** (according to NMR spectra) were isolated. The nitrone **14** (237 mg, 23 %) was obtained as an analytically pure, colourless solid, m. p. 152–155 °C.

$[\alpha]_D^{20}$ = –12 (c = 0.23, CHCl$_3$)

$C_{23}H_{31}NO_3$	calcd	C 74.76	H 8.46	N 3.79
(369.5)	found	C 74.32	H 8.43	N 3.71

IR (neat): ν = 2928 (s), 2902 (s), 2849 (s), 1449 (s, C=N), 1252 (s, N–O), 1114 (vs, C-O), 1045 (m), 925 (s, N–O), 752 (s), 687 (vs) cm^{-1}.

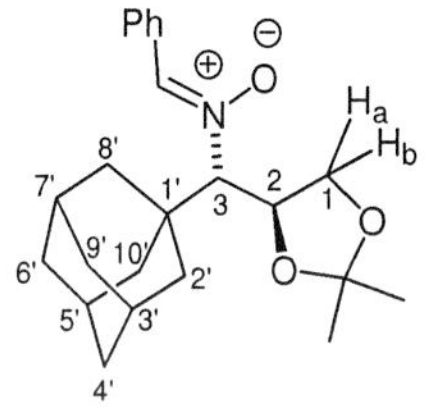

^{1}H NMR (300.1 MHz, CDCl$_3$): δ = 1.37 and 1.38 [2 s, 3 H each, C(CH$_3$)$_2$],1.68–1.70 (bs, 9 H), 2.01–2.07 (m, 6 H), 3.46 ("d", $^3J_{1,2}$ = 7.5 Hz,1 H, 3-H), 3.97 ("dd", $^2J_{1a,1b}$ = 8.9, $^3J_{1a,2}$ = 6.1 Hz, 1 H, 1-H$_a$), 4.17 ("dd", $^2J_{1a,1b}$ = 8.9, $^3J_{1b,2}$ = 6.2 Hz, 1 H, 1-H$_b$), 4.81 ("ddd", $^3J_{1a,2}$ = $^3J_{1b,2}$ = 6.1, $^3J_{2,3}$ = 7.5 Hz, 1 H, 2-H), 7.32 (s, 1 H, CH=N), 7.39–7.43 (m, 3 H of C$_6$H$_5$), 8.18–8.23 (m, 2 H of C$_6$H$_5$).

^{13}C NMR (75.5 MHz, CDCl$_3$): δ = 25.5 and 26.8 [2 q, C(CH$_3$)$_2$], 28.6 (d, C-3', C-5', C-7'), 36.1 (s, C-1'), 36.8 (t, C-4', C-6', C-9'), 39.6 (t, C-2', C-8', C-10'), 67.6 (d, C-3), 73.5 (d, C-2), 85.9 (t, C-1), 109.4 [s, C(CH$_3$)$_2$], 128.4 (d, o-, m-C of C$_6$H$_5$) 130.3 (d, p-C of C$_6$H$_5$), 130.3 (s, i-C of C$_6$H$_5$), 135.9 (d, C=N).

The assignment of signals was done by means of C,H COSY experiments.

Experiment 18 (AB 160)

(2S,3S)-3-(1-Adamantyl)-3-amino-1,2-O-isopropylidene-1,2-propanediol (15)

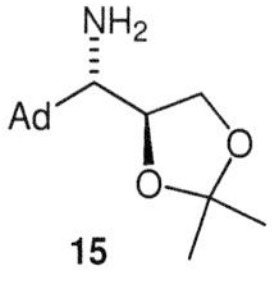

15

The removal of the N-benzyl and N-hydroxy groups was performed according to TP 4 without addition of di-tert-butyl dicarbonate:

N-Benzylhydroxylamine **7b** 85 mg, 0.23 mmol

Methanol 25 mL

Pd(OH)$_2$/C 20 mg was added

The crude product in form of a colourless solid was chromatographed on silica gel (5 g, 2 cm × 3 cm), elution with ethyl acetate/petroleum ether/Et$_3$N (50:50:2, v/v/v). From this, 30 mg (49 %) of the amine product **15** were isolated as an analytically pure, colourless solid, m. p. 39–41 °C.

$[\alpha]_D^{20}$ = 29 (c = 0.47, CHCl$_3$)

C$_{16}$H$_{27}$NO$_2$	calcd	C 72.41	H 10.25	N 5.28
(265.4)	found	C 72.39	H 10.18	N 5.24

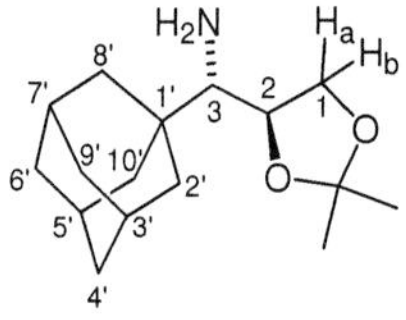

^{1}H NMR (300.1 MHz, CDCl$_3$): δ = 1.10 (bs, 2 H, NH$_2$), 1.35 and 1.43 [2 s, 3 H each, C(CH$_3$)$_2$], 1.48–1.52 and 1.61–1.74 (m each, 3 H and 9 H, 2'-H, 4'-H, 6'-H, 8'-H, 9'-H, 10'-H), 1.99 (bs, 3 H, 3'-H, 5'-H, 7'-H), 2.71 ("d", $^3J_{2,3}$ = 3.0 Hz, 1 H, 3-H), 3.87 ("dd", $^2J_{1a,1b}$ = 7.9, $^3J_{1a,2}$ = 7.8 Hz, 1 H, 1-H$_a$), 3.92 ("dd", $^2J_{1a,1b}$ = 8.0, $^3J_{1b,2}$ = 6.5 Hz, 1 H, 1-H$_b$), 4.29 ("ddd", $^3J_{1a,2}$ = 7.8, $^3J_{1b,2}$ = 6.5, $^3J_{2,3}$ = 3.4 Hz, 1 H, 2-H).

^{13}C NMR (75.5 MHz, CDCl$_3$): δ = 25.1 and 26.5 [2 q, C(CH$_3$)$_2$], 28.4 (d, C-3', C-5', C-7'), 35.1 (s, C-1'), 37.1 (t, C-4', C-6', C-9'), 39.3 (t, C-2', C-8', C-10'), 61.4 (d, C-3), 64.5 (t, C-1), 76.0 (d, C-2), 107.2 [s, C(CH$_3$)$_2$].

Experiment 19 (AB 148)

2,3-*O*-Cyclohexylidene-D-glyceraldehyde *N*-benzylnitrone (16)

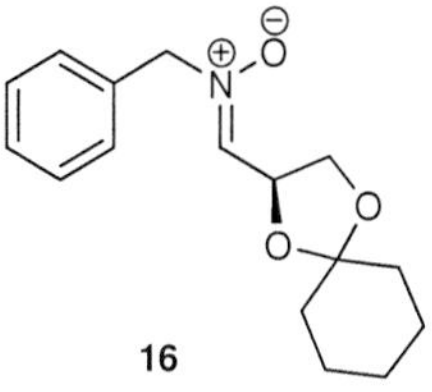

16

According to the literature,[56] 2,3-*O*-cyclohexylidene-D-glyceraldehyde (1.38 g, 8.1 mmol) and *N*-benzylhydroxylamine (1.00 g, 8.1 mmol) were dissolved in dry dichloromethane (20 mL), and magnesium sulfate (ca. 2 g) was added. The suspension was stirred overnight (17 h) at ambient temperature, and the solid was filtered off. The filtrate was evaporated and the residue recrystallized from ethyl acetate/petroleum ether to give the product **16** as a spectroscopically and analytically pure solid (1.65 g, 74 %), m. p. 97–100 °C (lit.:[147] 88–89 °C). The structure of **16** was confirmed by X-ray crystal structure analysis.

$[\alpha]_D^{20}$ = 79 (c = 1.28, EtOH), lit.:[147] $[\alpha]_D^{20}$ = 82.5 (c = 1.00, EtOH)

$C_{16}H_{21}NO_3$	calcd	C 69.79	H 7.69	N 5.09
(275.3)	found	C 69.75	H 7.72	N 5.05

IR (neat): ν = 2929 (s), 2852 (m, C=N–O), 1443 (s), 1273 (s, N–O), 1198 (s), 1168 (s), 1102 (vs, C–O), 1024 (s), 927 (vs, N–O), 706 (vs), 664 (s), 589 (s) cm^{-1}.

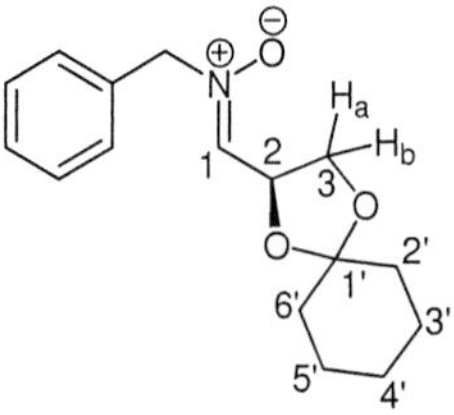

^{1}H NMR (300.1 MHz, CDCl$_3$): δ = 1.40 (m, 2 H, 4'-H), 1.58 (m, 8 H, 2'-H, 3'-H, 5'-H, 6'-H), 3.88 ("dd", $^3J_{2,3a}$ = 5.7, $^2J_{3a,3b}$ = 8.6 Hz, 1 H, 3-H$_a$), 4.38 ("dd", $^3J_{2,3b}$ = 7.0, $^2J_{3a,3b}$ = 8.6 Hz, 1 H, 3-H$_b$), 4.87 ("s", 2 H, A, B of CH$_A$H$_B$C$_6$H$_5$), 5.15 ("ddd", $^3J_{1,2}$ = 4.2, $^3J_{2,3a}$ = 5.7 Hz, $^3J_{2,3b}$ = 7.0, 1 H, 2-H), 6.86 ("d", $^3J_{1,2}$ = 4.2 Hz, 1 H, 1-H), 7.39–7.41 (m, 5 H, *o*-, *m*-, *p*-H of C$_6$H$_5$).

^{13}C NMR (75.5 MHz, CDCl$_3$): δ = 23.7 (t, C-4'), 23.9 and 25.0 (2 t, C-3' and C-5'), 34.3 and 35.9 (2 t, C-2', C-6'), 67.5 (t, C-3), 69.0 (t, CH$_2$C$_6$H$_5$), 71.7 (d, C-2), 110.6 (s, C-1'), 129.2 (d, p-C of C$_6$H$_5$), 129.0 and 129.4 (2 d, o-, m-C of C$_6$H$_5$), 132.1 (s, i-C of C$_6$H$_5$), 139.4 (d, C-1).

The assignment of signals was possible by means of DEPT and C,H COSY experiments. The NMR data were in accordance with those given in the literature.[147]

Experiment 20 (AB 149)

(2S,3S)-3-(1-Adamantyl)-3-(N-benzylhydroxylamino)-1,2-O-cyclohexylidene-1,2-propanediol (17)

17

The experiment was performed in accordance with TP 2:

2,3-O-Cyclohexylidene-D-glyceraldehyde N-benzylnitrone (**16**) 1.42 g, 5.2 mmol

Abs. diethyl ether 60 mL

Et$_2$AlCl ca. 5.2 mmol, 5.2 mL of ca. 1 M hexane solution

AdMgBr ca. 10.4 mmol

Stirring for 17 h at –60 °C (cryostate)

Quenched by saturated aqueous NH$_4$Cl solution (ca. 40 mL)

The crude product was isolated as a yellowish semi-solid, 2.50 g ("117 %"). Only one diastereomer of **17** was detected according to the ^{13}C NMR spectrum, as well as 10–15 % of the starting nitrone **16**. The crude material was purified by flash chromatography on silica gel (40 g, 14 cm × 3 cm column), elution with petroleum ether (ca. 100 mL) followed by elution with petroleum ether/ethyl acetate (80:20). The product was isolated as a colourless, analytically pure solid, m. p. 129–132 °C (1.29 g, 61 %).

$[\alpha]_D^{20}$ = 21.0 (c = 1.00, CHCl$_3$)

C$_{26}$H$_{37}$NO$_3$	calcd	C 75.87	H 9.06	N 3.40
(411.6)	found	C 75.41	H 9.35	N 3.31

IR (neat): ν = 3417 (w, OH), 2899 (s), 2843 (s, NH$_2$), 1447 (m), 1113 (vs, C–O), 1008 (s), 909 (s), 695 (s) cm^{-1}.

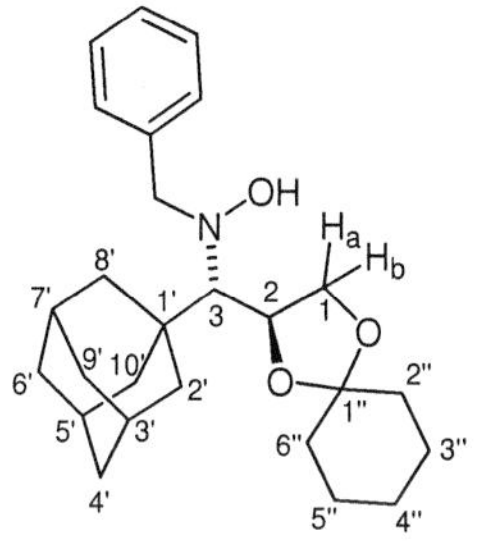

^{1}H NMR (300.1 MHz, CDCl$_3$): δ = 1.41 (m, 2 H, 4"), 1.55-1.70 (m, 14 H, 4'-H, 6'-H, 9'-H, 2"-H, 3"-H, 5"-H, 6"-H), 1.80 (bs, 6 H, 2'-H, 8'-H, 10'-H), 1.99 (bs, 3 H, 3'-H, 5'-H 7'-H), 2.57 ("d", $^3J_{2,3}$ = 6.3 Hz, 1 H, 3-H), 3.95 (A of AB, $^2J_{A,B}$ = 13.7 Hz, 1 H, CH$_A$H$_B$C$_6$H$_5$), 4.14 (dd, $^2J_{1a,1b}$ = 8.1, $^3J_{1a,2}$ = 8.8 Hz, 1 H, 1-H$_a$), 4.21 (B of AB, $^2J_{A,B}$ = 13.8 Hz, 1 H, CH$_A$H$_B$C$_6$H$_5$), 4.22 ("dd", $^2J_{1a,1b}$ = 8.1, $^3J_{1b,2}$ = 5.7 Hz, 1 H, 1-H$_b$), 4.45 ("ddd", $^3J_{1a,2}$ = 8.8, $^3J_{1b,2}$ = 5.7, $^3J_{2,3}$ = 6.3 Hz, 1 H, 2-H), 7.24–7.36 (m, 5 H, o-, m-, p-H of C$_6$H$_5$).

^{13}C NMR (75.5 MHz, CDCl$_3$): δ = 24.1 (t, C-3", C-5"), 25.3 (t, C-4"), 28.8 (d, C-3', C-5', C-7'), 35.1 and 36.6 (2 t, C-2", C-6"), 37.2 (t, C-4', C-6', C-9'), 38.0 (s, C-1'), 40.4 (t, C-2', C-8', C-10'), 64.6 (t, CH$_2$C$_6$H$_5$), 68.1 (t, C-1), 74.4 (d, C-3), 75.0 (d, C-2), 107.8 (s, C-1"), 127.1 (d, p-C of C$_6$H$_5$), 128.3 and 128.9 (2 d, o-, m-C of C$_6$H$_5$), 139.4 (s, i-C of C$_6$H$_5$).

Experiment 21 (AB 150)

(2S,3S)-3-(1-Adamantyl)-3-[N-(Z)-benzylidene-N-oxyamino]-1,2-cylohexylidene-1,2-propanediol (18)

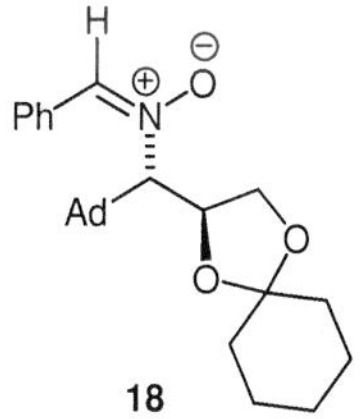

18

The hydrogenation was performed according to TP 4:

N-Benzylhydroxylamine **17** 3.85 g, 9.4 mmol

Methanol/THF mixture 35 mL:35 mL

Di-tert-Butyl dicarbonate 3.07 g, 14.1 mmol

Pd(OH)$_2$/C 120 mg

The hydrogenation was run for 3 days. The crude product, a colourless semi-solid was chromatographed on silica gel, elution with ethyl acetate/petroleum ether/triethylamine

(50:50:2, v/v/v). From this, 2.00 g ("52 %") of the starting material **17** in a mixture with **18** (according to NMR spectra) were isolated as well as 0.25 g (11 %) of the required spectroscopically pure *N*-Boc derivative **20**. The nitrone **18** was obtained as an analytically pure, colourless solid in 6 % yield (0.24 g), m. p. 148–151 °C.

$[\alpha]_D^{20}$ = 2.7 (c = 1.00, CHCl$_3$)

$C_{26}H_{35}NO_3$	calcd	C 76.25	H 8.61	N 3.42
(409.6)	found	C 76.30	H 8.66	N 3.23

IR (neat): ν = 2928 (s), 2902 (s), 2849 (s), 1449 (s, C=N), 1252 (s, N–O), 1114 (vs, C-O), 1045 (m), 925 (s, N–O), 752 (s), 687 (vs) cm^{-1}.

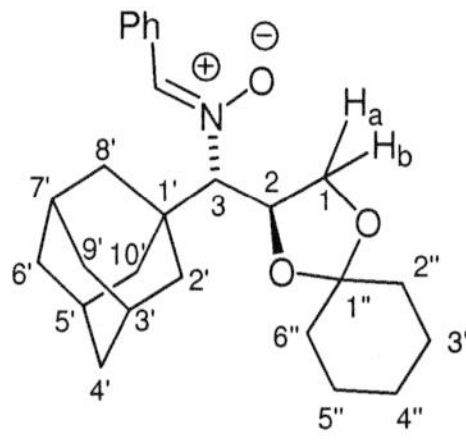

^{1}H NMR (500.1 MHz, CDCl$_3$): δ = 1.34–1.43 (m, 2 H), 1.54–1.70 (m, 18 H), 2.00–2.08 (m, 5 H), 3.45 ("d", $^3J_{1,2}$ = 7.7 Hz, 1 H, 3-H), 3.93 ("dd", $^2J_{1a,1b}$ = 8.8, $^3J_{1a,2}$ = 6.8 Hz, 1 H, 1-H$_a$), 4.15 ("dd", $^2J_{1a,1b}$ = 8.8, $^3J_{1b,2}$ = 6.1 Hz, 1 H, 1-H$_b$), 4.81 ("ddd", $^3J_{1a,2}$ = $^3J_{1b,2}$ = 6.1, $^3J_{2,3}$ = 7.7 Hz, 1 H, 2-H), 7.32 (s, 1 H, CH=N), 7.38–7.43 (m, 3 H of C$_6$H$_5$), 8.18–8.23 (m, 2 H of C$_6$H$_5$).

^{13}C NMR (75.5 MHz, CDCl$_3$): δ = 23.9 and 24.0 (2 t, C-3", C-5"), 25.2 (t, C-4"), 28.6 (d, C-3', C-5', C-7'), 36.1 (s, C-1'), 35.0 and 36.6 (2 t, C-2", C-6"), 36.8 (t, C-4', C-6', C-9'), 39.6 (t, C-2', C-8', C-10'), 67.4 (d, C-3), 73.1 (d, C-2), 86.0 (t, C-1), 110.1 (s, C-1"), 128.4 and 128.5 (2 d, *o*-, *m*-C of C$_6$H$_5$), 130.2 (s, *i*-C of C$_6$H$_5$), 130.3 (d, *p*-C of C$_6$H$_5$), 135.9 (d, C=N).

Signals were assigned by means of C,H COSY experiments.

Experiment 22 (AB 173)

(2S,3S)-3-(1-Adamantyl)-3-amino-1,2-O-cyclohexylidene-1,2-propanediol (19)

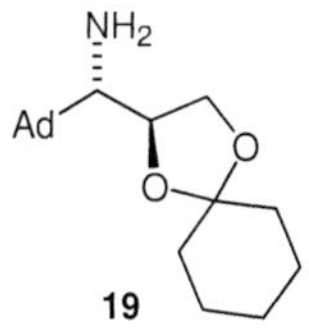

19

The removal of the *N*-benzyl and *N*-hydroxy groups was performed according to TP 4 without addition of di-*tert*-butyl dicarbonate:

N-Benzylhydroxylamine **17** 1.52 g, 3.7 mmol

Methanol/THF mixture 40 mL:10 mL

$Pd(OH)_2/C$ 160 mg was added

The crude product, a colourless semi-solid (1.13 g, "100 %"), was chromatographed on silica gel (40 g, 4 cm × 5 cm), elution with ethyl acetate/petroleum ether/triethylamine (50:50:2, v/v/v). From this, 0.99 g (87 %) of the amine product **19** were isolated as an analytically pure, colourless solid, m. p. 94–95 °C.

$[\alpha]_D^{20}$ = 21 (c = 1.2, $CHCl_3$)

$C_{19}H_{31}NO_2$	calcd	C 74.71	H 10.23	N 4.59
(305.5)	found	C 74.64	H 10.11	N 4.57

IR (neat): ν = 2901 (vs, b, NH_2), 2846 (vs, NH_2), 1446 (s), 1362 (s), 1165 (s), 1097 (vs, b, C-O), 1032 (vs, C–O), 940 (s,), 827 (s) cm^{-1}.

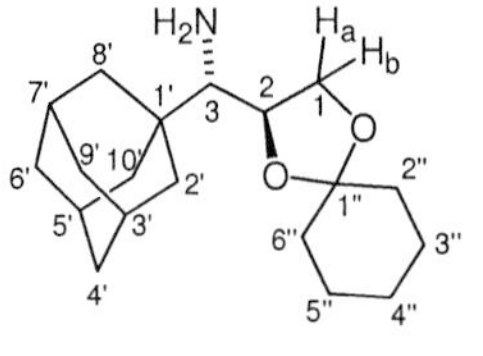

^1H NMR (300.1 MHz, $CDCl_3$): δ = 1.15 (bs, 2 H, NH_2), 1.40 (bs, 2 H, 4''-H), 1.49–1.74 (m, 20 H, 2'-H, 4'-H, 6'-H, 8'-H, 9'-H, 10'-H, 2''-H, 3''-H, 5''-H, 6''-H) ,1.98 (bs, 3 H, 3'-H, 5'-H, 7'-H), 2.70 ("d", $^3J_{2,3}$ = 3.6 Hz, 1 H, 3-H), 3.83 ("dd", $^2J_{1a,1b}$ = $^3J_{1a,2}$ = 8.0 Hz, 1 H, 1-H_a), 3.92 ("dd", $^2J_{1a,1b}$ = 7.9, $^3J_{1b,2}$ = 6.9 Hz, 1 H, 1-H_b), 4.26 ("ddd", $^3J_{1a,2}$ = 7.8, $^3J_{1b,2}$ = 6.9, $^3J_{2,3}$ = 3.7 Hz, 1 H, 2-H).

^{13}C NMR (75.5 MHz, CDCl$_3$): δ = 23.8 and 24.0 (2 t, C-3'', C-5''), 25.2 (t, C-4''), 28.4 (d, C-3', C-5', C-7'), 34.7 and 36.2 (2 t, C-2'', C-6''), 35.1 (s, C-1'), 37.1 (t, C-4', C-6', C-9'), 39.2 (t, C-2', C-8', C-10'), 61.4 (d, C-3), 64.4 (t, C-1), 75.6 (d, C-2), 107.8 (s, C-1'').

Experiment 23 (AB 192)

(2S,3S)-3-(1-Adamantyl)-3-(*tert*-butoxycarbonylamino)-1,2-*O*-cyclohexylidene-1,2-propanediol (20)

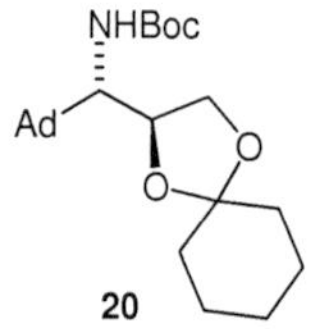

20

In accordance with a literature procedure,[114] the amine **19** (1.67 g, 5.5 mmol) was dissolved in acetonitrile (15 mL) and di-*tert*-butyl dicarbonate (1.20 g, 5.5 mmol) was added. The resulting solution was left with stirring for 20 h, while a colourless solid product precipitated. The solvent was removed *in vacuo* (18 mbar), and the crude product was recrystallized from ethyl acetate/petroleum ether. The *N*-Boc derivative **20** was isolated as a colourless, analytically pure solid, 1.69 g (76 %), m. p. 156–160 °C.

$[\alpha]_D^{20}$ = 10.9 (c = 1.66, CHCl$_3$)

C$_{24}$H$_{39}$NO$_4$	found	C 70.94	H 9.68	N 3.32
(405.6)	calcd	C 71.07	H 9.69	N 3.45

IR (neat): ν = 3420 (bw, N–H), 2930 (bs), 2851 (m), 1674 (vs, b, amide I), 1531 (vs, amide II), 1169 (vs, C–O), 1104 (vs, C–O), 1012 (m), 660 (m, b) cm^{-1}.

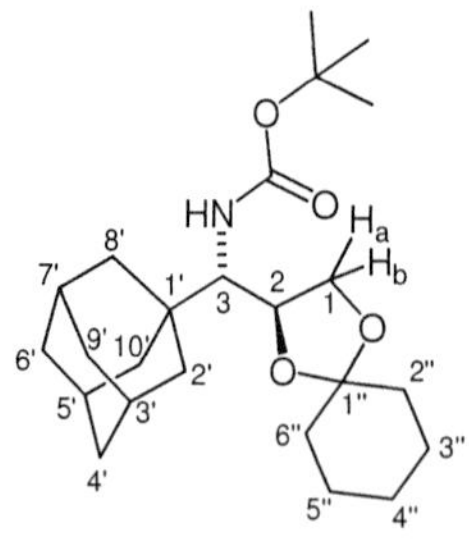

^{1}H NMR (300.1 MHz, CDCl$_3$): δ = 1.38–1.47 [m, 11 H, 4''-H and C(CH$_3$)$_3$], 1.57–1.73 (m, 20 H, 2'-H, 4'-H, 6'-H, 8'-H, 9'-H, 10'-H, 2''-H, 3''-H, 5''-H, 6''-H), 1.98 (bs, 3 H, 3'-H, 5'-H, 7'-H),

3.51 ("dd", $^3J_{2,3}$ = 6.8, $^3J_{3,NH}$ = 10.8 Hz, 1 H, 3-H), 3.68 ("t", $^2J_{1a,1b}$ = $^3J_{1a,2}$ = 8.1 Hz, 1 H, 1-H$_a$),
4.00 ("dd", $^2J_{1a,1b}$ = 8.4, $^3J_{1b,2}$ = 6.3 Hz, 1 H, 1-H$_b$), 4.19 ("dt", $^3J_{1b,2}$ = 6.3, $^3J_{1a,2}$ = $^3J_{2,3}$ = 8.1 Hz,
1 H, 2-H), 4.41 (d, $^3J_{3,NH}$ = 10.8 Hz, 1 H, NH).

^{13}C NMR (75.5 MHz, CDCl$_3$): δ = 23.9 and 24.0 (2 t, C-3", C-5"), 25.2 (t, C-4"), 28.3 [q,
C(CH$_3$)$_3$], 28.3 (d, C-3', C-5', C-7'), 35.4 and 36.1 (2 t, C-2", C-6"), 36.0 (s, C-1'), 37.0 (t,
C-4', C-6', C-9'), 39.0 (t, C-2', C-8', C-10'), 61.0 (d, C-3), 66.9 (t, C-1), 74.4 (d, C-2), 79.2 [s,
C(CH$_3$)$_3$], 109.6 (s, C-1"), 156.1 (s, C=O).

Experiment 24 (AB261)

(2S,3S)-3-(1-Adamantyl)-3-aminopropane-1,2-diol hydrochloride (21)

According to a known procedure,[148] (2S,3S)-3-(1-adamantyl)-3-amino-1,2-O-cyclohexylidene-
1,2-propanediol (**19**, 200 mg, 0.65 mmol) was dissolved in tetrahydrofuran (2 mL), treated
with concentrated hydrochloric acid (2 mL), and stirred for 16 h at ambient temperature. Then
the solvents were removed *in vacuo* (18 mbar) to give a colourless solid which after
recrystallization from methanol/diethyl ether mixture gave 606 mg (82 %) of the title 3-
aminodiol as a colourless, analytically pure solid, m. p. 235 °C (decomposition). The
structure was confirmed by X-ray crystal structure analysis.

$[\alpha]_D^{20}$ = 13 (c = 0.91, MeOH)

IR (neat): ν = 3052 (bm, OH), 2998 (vs), 2846 (s), 1602 (m), 1505 (s), 1029 (s) cm^{-1}.

C$_{13}$H$_{24}$ClNO$_2$	calcd	C 59.64	H 9.24	N 5.35	Cl 13.54
(261.8)	found	C 59.35	H 9.31	N 5.22	Cl 13.38

^{1}H NMR (300.1 MHz, MeOD): δ =1.70–1.81 (m, 12 H, 2'-H, 4'-H, 6'-H, 8'-H, 10'-H), 2.05 (bs,
3 H, 3'-H, 5'-H, 7'-H), 2.98 ("d", $^3J_{2,3}$ = 3.9 Hz, 1 H, 3-H), 3.86 ("d", $^3J_{1,2}$ = 3.5 Hz, 2 H, 1-H),

3.94 ("dd", $^3J_{1,2} = {}^3J_{2,3} = 3.7$ Hz, 1 H, 2-H).

^{13}C NMR (75.5 MHz, MeOD): δ = 29.5 (d, C-3', C-5', C-7'), 35.5 (s, C-1'), 37.5 (t, C-4', C-6', C-9'), 39.5 (t, C-2', C-8', C-10'), 65.2 (t, C-1), 67.2 (d, C-2), 68.5 (d, C-3).

Experiment 25 (AB 269)

(2S,3S)-3-(1-Adamantyl)-3-($tert$-butoxycarbonylamino)-1,2-O-cyclohexylidenepropane-1,2-diol (22)

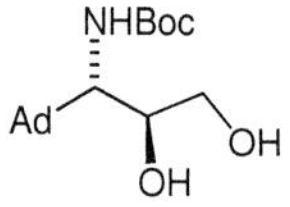

In analogy to the literature,[114] the amine hydrochloride **21** (100 mg, 0.38 mmol) was suspended in acetonitrile (3 mL) and treated with triethylamine (ca. 20 µL); after stirring for 1 h the pH of the solution became about 9. Then di-$tert$-butyl dicarbonate (83 mg, 0.38 mmol) was added, and the mixture was left with stirring for 21 h. The solvent was removed *in vacuo* (18 mbar), and the colourless solid was purified by filtration through silica gel (2.5 g silica gel, column 5 cm × 1.5 cm), elution with ethyl acetate/petroleum ether 50:50 (v/v). The product **22** was isolated as an analytically pure, colourless solid (98 mg, 79 %), m. p. 147–151 °C.

$[\alpha]_D^{20}$ = –5.9 (c = 0.56, CHCl$_3$)

IR (neat): ν = 2900 (m), 2886 (m) and 2846 (m) (OH, NH), 1694 (s), 1671 (vs, amide I), 1507 (s, NH), 1366 (s), 1158 (s), 1041 (s), 1012 (vs) cm^{-1}.

C$_{18}$H$_{31}$NO$_4$	calcd	C 66.43	H 9.60	N 4.30
(325.4)	found	C 66.25	H 9.61	N 4.13

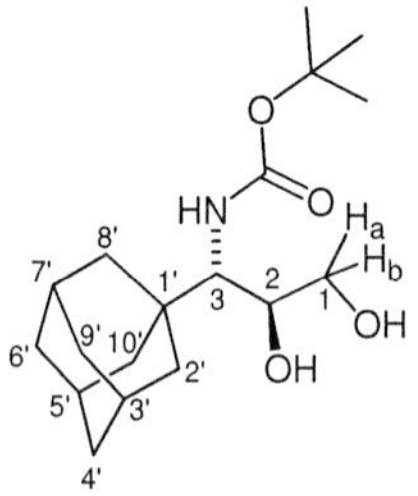

^{1}H NMR (500.1 MHz, CDCl$_3$): δ = 1.45 [s, 3 H, C(CH$_3$)$_3$], 1.66–1.72 (bm, 12 H, 2'-H, 4'-H, 6'-H, 8'-H, 10'-H), 2.00 (bs, 3 H, 3'-H, 5'-H, 7'-H), 3.17 ("dd", $^3J_{1a,OH}$ = 3.9, $^3J_{1b,OH}$ = 9.6 Hz,

1 H, 1-OH), 3.20 ("d", $^3J_{2,OH}$ = 8.9 Hz, 1 H, 2-OH), 3.27 ("dd", $^3J_{2,3}$ = 6.7, $^3J_{3,NH}$ = 9.3 Hz, 1 H, 3-H), 3.58 (m, 1 H, 1-H$_a$), 3.65 ("ddd", $^3J_{1b,OH}$ = 3.8, $^2J_{1a,1b}$ = 11.9, $^3J_{1b,2}$ = 4.2 Hz, 1 H, 1-H$_b$), 3.72 (m, 1 H, 2-H), 4.69 (d, $^3J_{3,NH}$ = 9.3 Hz, 1 H, NH).

^{13}C NMR (125.8 MHz, CDCl$_3$): δ = 28.4 [q, C(CH$_3$)$_3$], 28.4 (d, C-3', C-5', C-7'), 35.5 (s, C-1'), 36.8 (t, C-4', C-6', C-9'), 39.6 (t, C-2', C-8', C-10'), 61.8 (d, C-3), 64.7 (t, C-1), 71.1 (d, C-2), 80.3 [s, C(CH$_3$)$_3$], 157.8 (s, C=O).

The assignment of signals was done by means of C,H COSY experiments.

^{1}H NMR (500.1 MHz, DMSO-d$_6$): δ = 1.38 [s, 3 H, C(CH$_3$)$_3$], 1.50–1.69 (bm, 12 H, 2'-H, 4'-H, 6'-H, 8'-H, 9'-H, 10'-H), 1.90 (bs, 3 H, 3'-H, 5'-H, 7'-H), 3.11 ("dd", $^3J_{2,3}$ = 3.2, $^3J_{3,NH}$ = 7.2 Hz, 1 H, 3-H), 3.20 ("ddd", J = 1.3, J = 2.3, J = 5.0 Hz, 1 H, 1-H$_a$), 3.42–3.46 ("ddd", J = 1.5 , J = 2.8, J = 3.4 Hz, 1 H, 1-H$_b$), 3.53–3.58 (m, 1 H, 2-H), 4.38 (m, 2 H, 1-OH and 2-OH), 6.40 (d, $^3J_{3,NH}$ = 10.3 Hz, 1 H, NH).

^{13}C NMR (125.8 MHz, DMSO-d$_6$): δ = 27.9 [q, C(CH$_3$)$_3$], 28.3 (d, C-3', C-5', C-7'), 36.0 (s, C-1'), 36.7 (t, C-4', C-6', C-9'), 38.5 (t, C-2', C-8', C-10'), 61.0 (d, C-3), 62.5 (t, C-1), 70.5 (d, C-2), 77.2 [s, C(CH$_3$)$_3$], 156.0 (s, C=O).

The partial assignment of signals was done by means of C,H COSY experiments.

Experiment 26 (AB 222)

(S)-N-(tert-Butoxycarbonylamino)-1-adamantylglycine [(S)-23]

(S)-**23**

In analogy to the literature,[61,62] the diol **22** (370 mg,1.14 mmol) was dissolved in THF/water (60:40, 10 mL) and cooled to 0 °C. NaIO$_4$ (365 mg, 1.71 mmol) was added with vigorous stirring. The reacting suspension was left with stirring at 0–5 °C for 45 min, then it was filtered off and the solid was washed with diethyl ether (5 mL). The aqueous layer was extracted with diethyl ether (3 × 5 mL). The combined organic layers were washed with brine, dried over Na$_2$SO$_4$, and concentrated *in vacuo* (200 mbar, cold water bath). The residual colourless solid was dissolved in a mixture of *tert*-butanol (3 mL) and 2-methyl-2-butene

(5 mL). To this, an aqueous solution of $NaClO_2$ (154 mg, 1.71 mmol) and KH_2PO_4 (236 mg, 1.71 mmol) was added dropwise within 10 min at 0–10 °C. The resulting suspension was stirred for 1.5 h, followed by addition of the same amount of $NaClO_2$ and KH_2PO_4 at 0 °C. After stirring overnight (18 h) 3 N aqueous NaOH solution (ca. 5 mL) was added and volatiles were removed *in vacuo* (18 mbar). The residue was diluted with water (ca. 10 mL) and acidified by addition of 6 N hydrochloric acid (ca. 3 mL) to reach pH 3-4 (under cooling in an ice-bath). The solid formed was extracted with ethyl acetate (5 ×, 80 mL). The combined organic layers were washed with saturated aqueous Na_2SO_3 (2 × 5 mL) and brine (5 mL), dried over Na_2SO_4, and concentrated *in vacuo* (18 mbar) to give 357 mg (quant.) of a colourless solid. Recrystallization from ethyl acetate/petroleum ether gave 291 mg (82 %) of the spectroscopically and analytically pure product, m. p. >300 °C (decomposition).

$[\alpha]_D^{20}$ = 22.6 (c = 1.13, $CHCl_3$)

IR (neat): ν = 2972 (w), 2924 (m), 2901 (s), 2851 (m), 1713 (s, C=O of Boc), 1686 (s, amide I), 1645(s), 1506 (s, amide II), 1396 (m), 1153 (vs), 1041 (s), 1023 (m) cm^{-1}.

$C_{17}H_{27}NO_4$	calcd	C 65.99	H 8.80	N 4.53
(309.4)	found	C 66.05	H 8.66	N 4.51

^{1}H NMR (300.1 MHz, $CDCl_3$): δ = 1.45 [s, 9 H, $C(CH_3)_3$], 1.60–1.72 (bm, 12 H, 2'-H, 4'-H, 6'-H, 8'-H, 9'-H, 10'-H), 2.01 (bs, 3 H, 3'-H, 5'-H, 7'-H), 3.99 and 5.08 (2 d, $^3J_{2,NH}$ = 9.3 Hz, together 1 H, 2-H, rotamers).
The ^{1}H NMR spectrum corresponds to that of the (S)-enantiomer of N-(tert-butoxycarbonylamino)-1-adamantylglycine.[32]

^{13}C NMR (75.1 MHz, $CDCl_3$): δ = 28.2 [q, $C(CH_3)_3$], 28.3 (d, C-3', C-5', C-7'), 36.1 (s, C-1'), 36.6 (t, C-4', C-6', C-9'), 38.5 (t, C-2', C-8', C-10'), 62.4 (d, C-2), 79.9 [s, $C(CH_3)_3$], 155.7 (s, C=O of Boc), 176.3 (s, C=O).

At room temperature (295 K) the signals show satellite peaks at 64.5, 81.5 and 156.8 ppm, probably due to the presence of two rotamers. Recording of the NMR spectra at 375 K (102 °C) using deuterated dimethylsulfoxide as a solvent led to a coalescence of signals, which resulted in clear ^{1}H and ^{13}C NMR spectra:

^{1}H NMR (500.1 MHz, DMSO-d$_6$, 375 K): δ = 1.42 [s, 9 H, C(CH$_3$)$_3$], 1.60–1.71 (bm, 12 H, 2'-H, 4'-H, 6'-H, 8'-H, 9'-H, 10'-H), 1.98 (bs, 3 H, 3'-H, 5'-H, 7'-H), 3.72 (d, $^3J_{2,NH}$ = 9.2 Hz, 1 H, 2-H), 5.96 (d, $^3J_{2,NH}$ = 8.2 Hz, 1 H, NH).

^{13}C NMR (126 MHz, DMSO-d$_6$, 375 K): δ = 28.0 [q, C(CH$_3$)$_3$], 28.2 (d, C-3', C-5', C-7'), 35.1 (s, C-1'), 36.5 (t, C-4', C-6', C-9'), 38.5 (t, C-2', C-8', C-10'), 63.2 (d, C-2), 78.4 (s, [q, C(CH$_3$)$_3$], 155.4 (s, C=O of Boc), 171.8 (s, COOH).

Experiment 27 (AB 270)
(S)-1-Adamantylglycine hydrochloride [(S)-24]

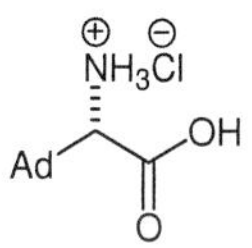

(S)-**24**

The N-Boc-protected amino acid (S)-**23** (50 mg, 0.16 mmol) was dissolved in THF (1 mL) and treated with 12 N HCl (1 mL). After stirring for 1 h a colourless solid formed. Stirring was continued for ca. 20 h, and solvents were removed *in vacuo* to give 40 mg (quant.) of the title compound as an analytically almost pure, colourless solid with a melting/decomposition range of 247–292 °C (lit.:[5] m. p. 236–240 °C, lit.:[149] m. p. 236–241 °C).

$[\alpha]_D^{20}$ = 20.9 (c = 0.93, MeOH) [lit.:[5] $[\alpha]_D$ = 16 (c = 0.50, MeOH); lit.:[149] $[\alpha]_D$ = 18 (c = 0.50, MeOH)].

IR (neat): ν = 3019 (m, NH$_2$), 2902 (vs, OH), 2848 (s), 2597 (m), 1507 (s, C=O), 1219 (s), 1030 (s) cm^{-1}.

For the X-ray and elemental analyses, a sample of [(S)-**24**] was recrystallized from methanol/diethyl ether mixture.

C₁₂H₂₀ClNO₂ ... $C_{12}H_{20}ClNO_2$ calcd   C 58.65   H 8.20   N 5.70   Cl 14.43

(245.7)

                          found   C 57.91   H 8.22   N 5.37   Cl 14.34

$C_{12}H_{20}ClNO_2 \cdot 1/2CH_3OH$ calcd   C 57.35   H 8.47   N 5.35   Cl 13.54

(261.6)

^{1}H NMR (300.1 MHz, MeOD): δ = 1.63–1.84 (m, 12 H, 2'-H, 4'-H, 6'-H, 8'-H, 9'-H, 10'-H), 2.06 (bs, 3 H, 3'-H, 5'-H, 7'-H), 3.52 (s, 1 H, 2-H).

^{13}C NMR (75.1 MHz, MeOD): δ = 29.3 (d, C-3', C-5', C-7'), 35.4 (s, C-3), 37.1 (t, C-4', C-6', C-9'), 39.0 (t, C-2', C-8', C-10'), 63.3 (d, C-2), 170.2 (C=O).

7.4   Experiments concerning Chapter 3

Experiment 28 (AB 105)

Benzophenone N-(p-anisylmethyl)-imine (25)

25

Following a procedure described in the literature,[95] benzophenone (0.40 g, 2.19 mmol) and p-methoxybenzylamine (0.30 g, 2.2 mmol) were dissolved in dry toluene (20 mL). ZnBr₂ (11 mg) was added, and the resulting solution under nitrogen was heated under reflux over molecular sieves (4 Å) for two days (42 h). Then it was cooled to room temperature and passed through a celite pad (2 cm). The adsorbent was washed with toluene (2 × 10 mL), and the filtrate was concentrated under reduced pressure. The recrystallization of a residual oil from petroleum ether gave 0.38 g (57 %, lit.:[95] 72 %) of a colourless, analytically pure solid (m. p. 64–66 °C; lit.:[95] 76 °C).

$C_{21}H_{19}NO$ calcd   C 83.69   H 6.35   N 4.65

(301.2)   found   C 83.31   H 6.40   N 4.62

IR (neat): ν = 1612 (m), 1575 (m), 1509 (s, C=N), 1444 (s), 1313 (m), 1284 (m), 1237 (vs), 1173 (s), 1031 (s, C-O), 1015 (s), 850 (m), 824 (s), 800 (s), 777 (s), 760 (s), 697 (vs), 638 (s) cm^{-1}.

^{1}H NMR (300.1 MHz, $CDCl_3$): δ = 3.78 (s, 3 H, OCH_3), 4.54 ("s", 2 H, $CH_2C_6H_4$), 6.85-7.67 (m, 14 H, C_6H_4 and 2 C_6H_5).

^{13}C NMR (75.5 MHz, $CDCl_3$): δ = 55.3 (q, CH_3O), 56.9 (t, $CH_2C_6H_4$), 113.8 (d, C-3 and C-5), 127.8 (d), 128.0 (d), 128.5 (d), 128.5 (d, p-C of C_6H_5), 128.6 (d), 128.7 (d), 130.0 (d, p-C of C_6H_5), 132.8 (s, C-1), 136.8 (s, i-C of C_6H_5), 139.8 (s, i-C of C_6H_5), 158.3 (s, C-4), 168.5 (s, C=N).

The ^{1}H and ^{13}C NMR data were in accordance with the literature data.[95]

Experiment 29 (AB 109)
(1,1-Diphenyl-but-3-enyl)-(4-methoxy-benzyl)-amine (26)

26

Typical procedure for the allylation of N-protected benzophenone and acetone imines (TP 6)[19]

A round-bottomed, two-arm flask was charged with benzophenone N-(p-anisylmethyl)-imine **25** (0.70 g, 2.34 mmol) and absolute ether (15 mL) and cooled to 0 °C. Then allylmagnesium bromide (ca. 7.02 mmol, 3.3 mL of ca. 2.1 M ethereal solution) was added, and the resulting solution was stirred overnight (then poured into a cooled saturated solution of aqueous $NaHCO_3$ (10 mL). The aqueous layer was separated and extracted with ethyl acetate (3 × 10 mL). The combined organic layers were washed with brine (10 mL) and dried over Na_2SO_4. Removal of the solvent *in vacuo* (9 mbar) afforded 0.76 g (95 %) of a colourless solid. Flash chromatography on SiO_2 (20 g) (eluting with ethyl acetate/petroleum ether/Et_3N 20:80:2, v/v) gave 0.75 g (93 %) of the title product **26** as a spectroscopically and analytically pure, colourless solid (m. p. 85–90 °C).

$C_{24}H_{25}NO$ calcd   C 83.93   H 7.34   N 4.08

(343.2)   found   C 83.83   H 7.37   N 4.06

IR (neat): v = 2999 (w), 2950 (w, NH), 2831 (w), 1609 (w), 1511 (s, Ar), 1439 (s), 1303 (m), 1241 (s), 1176 (s), 1115 (m), 1015 (s, C-O), 1001 (m), 922 (m), 837 (m), 810 (m), 796 (m), 752 (s), 725 (s), 696 (m), 661 (m), 624 (m), 558 (m) cm^{-1}.

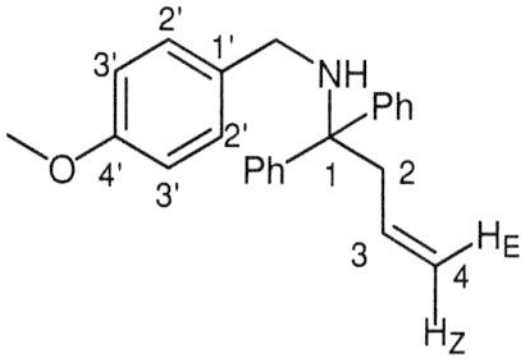

^{1}H NMR (300.1 MHz, CDCl$_3$): δ = 1.80 (bs, 1 H, NH), 3.15 ("dt", $^3J_{2,3}$ = 7.0, $^5J_{2,4HE}$ = $^5J_{2,4HZ}$ = 1.3 Hz, 2 H, 2-H), 3.37 ("s", 2 H, CH$_2$Ar), 3.78 (s, 3 H, CH$_3$O), 5.02 ("dd", $^3J_{3,4Z}$ = 10.1, $^2J_{4E,4Z}$ = 3.3 Hz, 1 H, 4-H$_Z$), 5.11 ("dd", $^3J_{3,4E}$ = 17.2, $^2J_{4E,4Z}$ = 3.6 Hz, 1 H, 4-H$_E$), 5.54 ("ddd", $^3J_{2,3}$ = 7.0, $^3J_{3,4HZ}$ = 10.1, $^3J_{3,4HE}$ = 17.1 Hz, 1 H, 3-H), 6.83 ("d", $^3J_{2',3'}$ = 9.0 Hz, 2 H, 3'-H), 7.14–7.31 (m, 8 H), 7.41–7.45 (m, 4 H).

^{13}C NMR (75.5 MHz, CDCl$_3$): δ = 41.1 (t, C-2), 45.9 (t, CH$_2$C$_6$H$_4$), 55.3 (q, CH$_3$O), 64.3 (s, C-1), 113.7 (d, C-3'), 118.0 (t, C-4), 126.3 (d, p-C of C$_6$H$_5$), 127.2 and 128.0 (2 d, o-, m-C of C$_6$H$_5$), 129.3 (d, C-2'), 133.3 (s, C-1'), 134.1 (d, C-3), 146.7 (2 s, i-C of C$_6$H$_5$), 158.6 (s, C-4').

4-(4-Methoxy-benzylamino)-4,4-diphenyl-butane-1,2-diol (27)

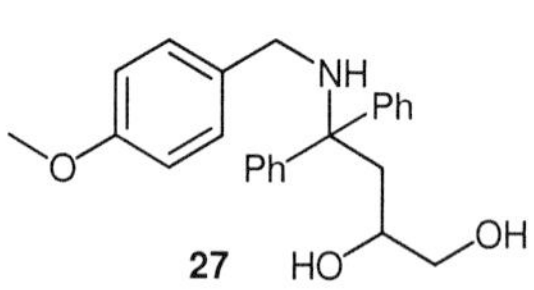

Experiment 30 (AB 129)

Preparation of the diol **27** by oxidation with osmium tetraoxide^{98}

To a solution of the allyl derivative **26** (200 mg, 0.58 mmol) in *tert*-butanol/THF (5 mL/1 mL) an aqueous of 4-methylmorpholine-4-oxide (NMO) (940 mg, 1.2 eq, 10 wt % solution) was added as well as a *tert*-butanol solution of OsO_4 (0.58 mL, 0.1 eq, 2.5 wt %). The resulting yellow solution was stirred for 6 h. Then ice (ca. 2 g) was added followed by addition of dry sodium sulfite (0.15 g), and the solution was stirred for 1 h (it became brown). After diluting with water (20 mL) the aqueous layer was separated and extracted with ethyl acetate (3 × 15 mL). The combined organic layers were dried over Na_2SO_4 and concentrated *in vacuo* (17 mbar) to give 230 mg ("105 %") of a colourless oil. It was purified by column chromatography on silica gel (20 g), eluting with ethyl acetate/petroleum ether 1:1 (v/v). The yield of colourless, analytically pure solid of **27** was 85 mg (39 %), m. p. 140–144 °C.

$C_{24}H_{27}NO_3$	calcd	C 76.36	H 7.21	N 3.71
(377.2)	found	C 75.92	H 7.25	N 3.68

IR (neat): ν = 3547 (w), 3265 (w), 2866 (w), 1610 (m), 1510 (m), 1465 (m), 1444 (m), 1244 (s, C-O), 1183 (m), 1129 (m), 1094 (m), 1021 (m), 945 (m), 916 (m), 859 (m), 822 (s), 812 (s), 760 (s), 702 (vs), 637 (m) cm^{-1}.

^{1}H NMR (500.1 MHz, CDCl$_3$): δ = 1.98 (dd, $^3J_{2,3a}$ = 13.9, $^2J_{3a,3b}$ = 1.9 Hz, 1 H, 3-H$_a$), 2.18 (bs, 1 H, NH), 2.85 (dd, $^3J_{2,3b}$ = 13.9, $^2J_{3a,3b}$ = 3.0 Hz, 1 H, 3-H$_a$), 3.09 (A of AB, $J_{A,B}$ = 11.6 Hz, 1 H, CH$_A$H$_B$C$_6$H$_4$), 3.38 (m, 1 H, H-2), 3.45 (m, 1 H, 1-H$_a$), 3.59 (m, 1 H, 1-H$_b$), 3.66 (B of AB, $J_{A,B}$ = 11.6 Hz, 1 H, CH$_A$H$_B$C$_6$H$_4$), 3.79 (s, 3 H, CH$_3$O), 6.83 ("d", $^3J_{2',3'}$ = 8.7 Hz, C-3'), 7.09 ("d", $^3J_{2',3'}$ = 8.7 Hz, C-2'), 7.21–7.39 (m, 10 H, 2 C$_6$H$_5$).

^{13}C NMR (75.5 MHz, CDCl$_3$): δ = 36.8 (t, C-3); 46.4 (t, CH$_2$C$_6$H$_4$); 55.3 (q, CH$_3$O); 66.1 (s, C-4); 66.9 (t, C-1); 69.5 (d, C-2); 114.1 (d, C-3'); 126.7 (d, *p*-C of C$_6$H$_5$); 126.9 (d, *p*-C of C$_6$H$_5$); 127.1 (d), 127.2 (d), 128.2 (d) and 128.5 (4 d, *o*-, *m*-C of C$_6$H$_5$), 129.7 (d, C-2'), 131.1 (s, C-1'), 146.1 and 146.4 (2 s, *i*-C of 2 C$_6$H$_5$), 159.0 (s, C-4').

Experiment 31 (AB 133)

Preparation of the diol **27** by oxidation with potassium osmate dihydrate[99]

The suspension of the allyl derivative **26** (200 mg, 0.58 mmol), potassium hexacyanoferrate $K_3Fe(CN)_6$ (III) (575 mg, 1.75 mmol) and potassium osmate dihydrate $K_2OsO_2(OH)_4$ (0.058 mmol, 21 mg) in a *tert*-butanol/water (1:1) mixture (15 mL) was cooled to 0 °C and potassium carbonate (242 mg, 1.75 mmol) was added. The suspension was left with stirring for 5 days, then sodium sulfite (400 mg) was added, and the mixture was stirred for an additional hour. The organic upper layer was separated, and the aqueous layer was extracted with ethyl acetate (3 × 10 mL). The combined organic layers were dried over Na_2SO_4 and concentrated *in vacuo* (17 mbar). Recrystallization of the residual oil from diethyl ether gave 162 mg (74 %) of **27** as a colourless, spectroscopically pure solid (m. p. 137–139 °C).

$C_{24}H_{27}NO_3$	calcd	C 76.36	H 7.21	N 3.71
(377.2)	found	C 74.43	H 7.05	N 3.59

IR, ^{1}H and ^{13}C NMR spectra see above (experiment 30, AB 129).

Experiment 32 (AB 136)

3-(4-Methoxybenzylamino)-3,3-diphenylpropionic acid (28)

Following a known procedure,[61,62] the diol **27** (500 mg, 1.33 mmol) was dissolved in methanol/water (1:1, 20 mL), cooled to 0 °C, and $NaIO_4$ (342 mg, 1.60 mmol) was added under vigorous stirring. The reaction suspension was left with stirring for 20 min at ambient temperature, then diluted with water (10 mL) and extracted with diethyl ether (3 × 10 mL). The combined organic layers were washed with brine, dried over Na_2SO_4, and concentrated *in vacuo* (100 mbar, cold water bath). The residual colourless oil was dissolved in a *tert*-butanol/2-methyl-2-butene mixture (5 mL/3 mL). To this a solution of $NaClO_2$ (240 mg, 2.66 mmol,) and KH_2PO_4 (362 mg, 2.66 mmol) in water (5 mL) was added dropwise within 30 min at 0 °C. The resulting suspension was stirred overnight (17h); 20 % aqueous NaOH solution was then added until pH>9, and the volatiles were removed *in vacuo* (50 mbar). The residual aqueous solution was washed with diethyl ether (2 × 10 mL) and acidified to pH 2-3 by addition of concentrated HCl. The solid formed was extracted with dichloromethane (5 × 10 mL). The combined organic solutes were washed with saturated aqueous Na_2SO_3 solution, dried over Na_2SO_4 and concentrated *in vacuo* (16 mbar) to give 260 mg ("53 %") of a colourless solid which was soluble neither in water nor methanol or any other organic solvent. It was not possible to characterize the substance by NMR spectroscopy.

Experiment 33 (AB 240, 247)
**Methyl 3-(4-methoxybenzylamino)-3,3-
diphenylpropanoate (29)**

29

The ethereal solution of diazomethane was prepared as follows:[150]

The condenser was filled with dry ice, then isopropanol was added slowly until the cold-finger was about one-third full. The solution of potassium hydroxide (10 g) in water (16 mL) as well as ether (16 mL) and 2-ethoxyethanol (28 mL) were added into the reaction vessel. A 100 mL receiving flask was attached to the condenser and the receiver was cooled by a dry ice/isopropanol bath. The side-arm of the condenser was connected to the diazomethane trap filled with glacial acetic acid.

A separatory funnel was placed over the reaction vessel and charged with a solution of Diazald® (5.0 g, 23 mmol) in ether (45 mL). The reaction vessel was warmed to 65 °C and the Diazald® solution was added dropwise over a period of 20 min. When Diazald® has been used up, the reaction vessel was filled with additional 20 mL of ether and the distillation was continued until the distillate was colourless.

The ethereal solution prepared in this way contained 17 to 20 mmol of diazomethane (0.20 to 0.25 M solution). It was stored in a fridge over solid KOH.

Based on a known procedure,[61,62] 4-(4-methoxybenzylamino)-4,4-diphenylbutane-1,2-diol (200 mg, 0.53 mmol) was dissolved in THF/water mixture (6:4, 10 mL), cooled to 0 °C, and $NaIO_4$ (136 mg, 0.64 mmol) was added under vigorous stirring. The reaction was left with stirring at 0–5 °C for 2 h, then the organic layer was separated and the aqueous layer was extracted with diethyl ether (3 ×, 15 mL). The combined organic layers were washed with brine, dried over Na_2SO_4, and concentrated *in vacuo* (200 mbar, water-cooling). The residual colourless oil was dissolved in a mixture of *tert*-butanol (4 mL) and 2-methyl-2-butene (3 mL). To this mixture a solution of $NaClO_2$ (144 mg 1.60 mmol) and NaH_2PO_4 (220 mg 1.60 mmol) in water (4 mL) was added dropwise within 30 min at 0 °C. The resulting suspension was stirred overnight (18 h) followed by addition of $NaClO_2$ (36 mg 0.5 mmol,) and NaH_2PO_4

(55 mg, 0.5 mmol) in water. Stirring was continued for 1 h, and the volatiles were removed *in vacuo* (0.1 mbar) to dryness. The colourless solid left was suspended in dry dichloromethane (20 mL) and diazomethane (solution in diethyl ether) was added, until the yellow colour persisted. The suspension was stirred for an additional 30 min, then water (ca. 15 mL) was added. The layers were separated, and the aqueous phase was extracted with dichloromethane (4 ×, 50 mL). Drying and concentration of the combined organic solutes gave 218 mg of a colourless oil. The ^{1}H NMR spectrum showed a complex mixture of compounds. The crude product was purified by column chromatography on silica gel (6 g, 6 cm × 2.5 cm), eluting with ethyl acetate/petroleum ether (50:50, v/v). A colourless oil (115 mg) was isolated; NMR analysis showed that the separation had not been effective. The column-chromatographic purification was repeated (3 g SiO$_2$, 17 cm × 1.5 cm), eluting with ethyl acetate/petroleum ether, 5:95 (v/v). The first fraction (40 mg, a colourless oil) represented an undefined mixture of compounds. The title compound was isolated as a second fraction (55 mg, 28 %), a colourless, analytically almost pure solid, m. p. 117–118 °C.

C$_{24}$H$_{25}$NO$_3$	calcd	C 76.77	H 6.71	N 3.73
(375.5)	found	C 76.19	H 6.74	N 3.65

IR (neat): ν = 3324 (bw, NH), 3004 (w), 1719 (vs, C=O), 1204 (vs, C-O, ester), 1167 (vs, C-O, ether), 1008 (s) cm^{-1}.

^{1}H NMR (300 MHz, CDCl$_3$): δ = 2.70 (bs, 1 H, NH), 3.45–3.47 (bm, 7 H, CO$_2$CH$_3$, 2-H, CH$_2$C$_6$H$_4$), 3.79 (s, 3 H, OCH$_3$), 6.85 (d, $^3J_{2',3'}$ = 8.5 Hz, 2 H, 2'-H), 7.18–7.20 (m, 2 H), 7.23–7.33 (m, 6 H of 2 C$_6$H$_5$), 7.41–7.44 (m, 4 H of 2 C$_6$H$_5$).

^{13}C NMR (75 MHz, CDCl$_3$): δ = 40.3 (t, C-2), 46.4 (t, CH$_2$C$_6$H$_4$), 51.4 (q, CO$_2$CH$_3$), 55.3 (q, CH$_3$O), 64.0 (s, C-3), 113.7 (d, C-3'), 126.6 (d, C-2'), 126.9 and 128.1 (2 d, *o*-, *m*-C of C$_6$H$_5$), 129.4 (d), 133.1 (s, C-1'), 145.8 (s, *i*-C of C$_6$H$_5$), 158.5 (s, C-4'), 171.5 (s, C=O).

Experiment 34 (AB 167)
Benzophenone oxime (30)

$$N-OH$$
$$Ph \quad Ph$$
30

As described in a literature procedure,[104] benzophenone (20.0 g, 110 mmol), hydroxylamine hydrochloride (7.64 g, 110 mmol) and pyridine (8.68 g, 110 mmol) were dissolved in dry ethanol (150 mL) and heated under reflux for 2 h. The solvent was removed *in vacuo* (20 mbar), and the residual oil was precipitated by addition of 25 % aqueous ammonia (ca. 150 mL). The colourless solid formed was filtered off and recrystallized from methanol (ca. 200 mL) to give 14.16 g of the product. More product (1.93 g) was isolated by repeated crystallisation of the mother liquid. The total yield was 16.09 g (74 %, lit.:[104] 95 %); m. p. 137–144 °C (lit.:[104] 143.5–144.5 °C).

$C_{13}H_{11}NO$	calcd	C 79.16	H 5.62	N 7.10
(197.2)	found	C 79.01	H 5.64	N 7.15

IR (neat): ν = 3247 (bw, OH), 2886 (bw), 1492 (bs), 1328 (bs), 994 (vs, N-O), 918 (vs), 765 (vs), 694 (vs), 567 (vs) cm^{-1}.

^{1}H NMR (300.1 MHz, CDCl$_3$): δ = 7.28–7.38 (m, 3 H), 7.40–7.50 (m, 7 H), 9.20 (bs, 1 H, OH).

^{13}C NMR (75.5 MHz, CDCl$_3$): δ = 127.9, 128.2, 128.3, 128.4, 129.2 and 130.1 (6 d, *o*-, *m*-C of 2 C$_6$H$_5$), 129.1 (d, *p*-C of C$_6$H$_5$), 129.5 (d, *p*-C of C$_6$H$_5$), 132.7 (s, *i*-C of C$_6$H$_5$), 136.2 (s, *i*-C of C$_6$H$_5$), 157.9 (s, C=N).

Experiment 35 (AB 193)
N-Diphenylphosphinoyl benzophenone imine (31)
and
O-(diphenylphosphinyl) benzophenone oxime (32)

31   **32**

Following a known procedure,[151,152] benzophenone oxime (**30**) (1.00 g, 5.1 mmol) and triethylamine (0.62 g, 6.1 mmol) were dissolved in THF (10 mL), and the solution was cooled to 0 °C. Chlorodiphenyl phosphine (also known as diphenylphosphinous chloride) (1.18 g, 5.3 mmol) was added dropwise at this temperature under nitrogen. The resulting mixture was stirred for 20 h, then triethylamine (0.30 g, 3.0 mmol) as well as chlorodiphenyl phosphine (0.50 g, 2.2 mmol) were added again, and the suspension was left with stirring for 7 h under the same conditions, followed by the hydrolysis with water (30 mL). The aqueous layer was extracted with ethyl acetate (4 ×, 100 mL), the combined organic layers were washed with water, dried, and concentrated *in vacuo* (18 mbar) to give 0.63 g of a brown oil which was purified by column chromatography on silica gel (20 g, 8 cm × 3 cm), eluting with a mixture petroleum ether/ethyl acetate (70:30 to 50 : 50, v/v).

1st fraction: a colourless solid, 0.11 g (11 %), starting material, benzophenone oxime (**30**) (according to NMR data).

2nd fraction: a colourless solid, 40 mg (2 %), *O*-(diphenylphosphinyl) benzophenone oxime (**32**), m. p. 183–185 °C (lit.:[151] 185 °C).

NMR data of the oxime **32** (not detected in the literature data):

^{13}C NMR (75.5 MHz, CDCl$_3$): δ = 127.9, 128.2, 128.3, 128.3, 128.5, 128.9, 129.1, 129.5, 129.6, 130.7, 131.7, 131.9, 132.1, 132.2, 132.3, 165.7, 165.9, 180.3.

3rd fraction: a yellowish sticky foam, 0.82 g (42 %), which after recrystallisation from diethyl ether gave 0.61 g (31 %) of analytically pure *N*-diphenylphosphinoyl benzophenone imine (**31**) as a colourless solid, m. p. 123–126 °C (lit.:[151] 124 °C).

C$_{25}$H$_{20}$NOP	calcd	C 78.73	H 5.29	N 3.67	P 8.12
(381.4)	found	C 78.73	H 5.33	N 3.59	P 7.97

IR (neat): ν = 1622 (s, C=N), 1573 (m, P-Ph), 1436 (s), 1265 (m, P=O), 1210 (vs), 1105 (s), 844 (s), 696 (vs) cm^{-1}.

^{13}C NMR (75.5 MHz, CDCl$_3$): δ = 127.9, 128.2, 128.4, 129.6, 131.2, 131.3, 131.3, 131.6, 131.8 (9 d, *o*-, *m*-, *p*-C of 4 C$_6$H$_5$), 134.1 (s, *i*-C of C$_6$H$_5$), 136.1 (s, *i*-C of C$_6$H$_5$), 138.7 (s, *i*-C of C$_6$H$_5$), 138.9 (s, *i*-C of C$_6$H$_5$), 181.8 (s, C=N).

Experiment 36 (AB 195)

N-(Allyldiphenylmethyl)diphenylphosphinic amide (33)

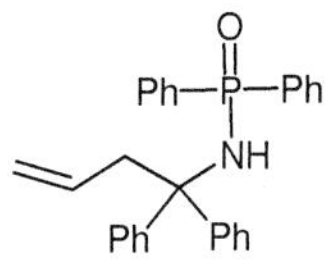

33

The reaction was carried out according to TP 6:

N-Diphenylphosphinoyl benzophenone imine (**31**) 0.10 g, 0.26 mmol
THF absolute 5 mL
Allylmagnesium bromide in THF (ca. 2 M) ca. 0.31 mmol, 0.15 mL

The resulting suspension was stirred for 20 h and hydrolyzed by addition of saturated aqueous NH_4Cl solution (10 mL). The aqueous layer was separated and extracted with ethyl acetate (5 ×, 50 mL). The combined organic layers were dried and concentrated *in vacuo* to yield 0.174 g of a colourless oil, which was crystallized from ethyl acetate/petroleum ether to give 0.91 g (82 %) of a colourless, analytically pure solid, m. p. 147–150 °C.

$C_{28}H_{26}NOP$	calcd	C 79.41	H 6.19	N 3.31	P 7.31
(423.5)	found	C 79.28	H 6.18	N 3.24	P 7.06

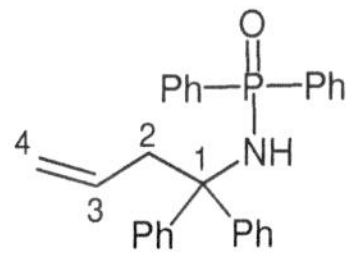

^{1}H NMR (300.1 MHz, $CDCl_3$): δ = 3.43 ("d", J = 5.5 Hz, 2 H, 2-H), 3.84 (d, J = 3.4 Hz, 1 H), 5.07 (dd, J = 4.1, J = 8.3 Hz, 1 H), 5.33 (m, 1 H), 7.06–7.13 (m, 6 H), 7.24–7.35 (m, 10 H), 7.65–7.73 (m, 4 H).

^{13}C NMR (75.5 MHz, $CDCl_3$): δ = 46.7 (t, C-2), 65.0 (s, C-1), 120.7 (t, C-4), 126.7 (d), 127.6 (d), 128.0 (d), 128.1 (d), 128.4 (d), 130.9 (d), 131.4 (d), 131.6 (d), 133.5 (s), 133.9 (s), 135.2 (d, C-3), 144.1 (s).

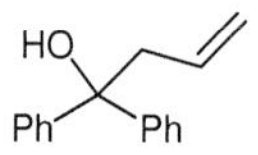

Experiment 37 (AB 182)
1,1-Diphenylbut-3-en-1-ol (34)

34

The reaction was carried out according to TP 6:

Benzophenone 5.00 g, 27.5 mmol
Abs. diethyl ether 100 mL
Allylmagnesium bromide ca. 34 mmol, 17 mL of ca. 2 M THF solution

The resulting solution was stirred for 3 h and then poured onto an ice/NH_4Cl mixture (ca. 100 mL). The aqueous layer was separated and extracted with diethyl ether (3 × 50 mL); the organic phases were washed with brine (80 mL), dried over Na_2SO_4 and concentrated *in vacuo* (400 mbar to 0.1 mbar) to give 6.07 g (99 %) of the product **34** as a spectroscopically pure, colourless oil.

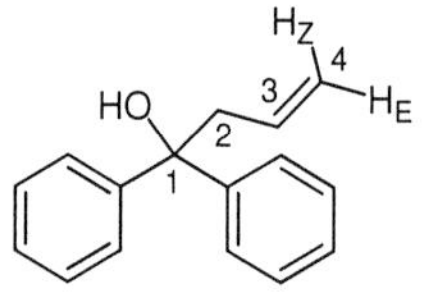

^{1}H NMR (300.1 MHz, $CDCl_3$): δ = 3.08 ("d", $^3J_{2,3}$ = 7.2 Hz, 2 H, 2-H), 5.17 ("d", $^3J_{3,4Z}$ = 10.1 Hz, 1 H, 4-H_Z), 5.24 ("d", $^3J_{3,4E}$ = 17.2 Hz, 1 H, 4-H_E), 5.68 (m, 1 H, 3-H), 7.19–7.24 (m, 2 H, *p*-H of 2 C_6H_5), 7.28–7.34 (m, 4 H), 7.43–7.48 (m, 4 H).

^{13}C NMR (75 MHz, $CDCl_3$): δ = 46.7 (t, C-2), 76.9 (s, C-1), 120.6 (t, C-4), 126.0 (d, *o*-C of 2 C_6H_5), 126.9 (d, *p*-C of 2 C_6H_5), (d, *m*-C of 2 C_6H_5),133.5 (d, C-3), 146.5 (s, *i*-C of 2 C_6H_5).

Assignment of the signals was done by means of C,H COSY experiments.

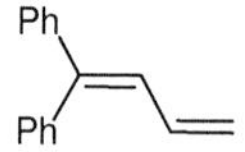

Experiment 38 (AB 183)
1,1-Diphenyl-1,3-butadiene (35)

35

According to the literature,[153] the alcohol **34** (0.50 g, 2.25 mmol) was dissolved in acetonitrile (2 mL) and conc. sulphuric acid (25 μl) was added. The solution was left with stirring for 20 h at ambient temperature, then quenched by addition of a few drops of triethylamine, and volatiles were removed under reduced pressure (18 mbar). The residual oil was purified on silica gel (20 g, eluting with ethyl acetate/petroleum ether 5:95, v/v). A colourless oil (0.19 g, 41 %) was isolated, the structure of which was shown by NMR data to be that of the diene **35**, as a comparison with literature data[107] showed.

^{1}H NMR (250 MHz, CDCl$_3$): δ = 5.11 (dd, $^2J_{4H_Z,4H_E}$ = 1.8, $^3J_{3,4H_Z}$ = 10.1 Hz, 1 H, 4-H$_Z$), 5.37 (dd, $^2J_{4H_Z,4H_E}$ = 1.8, $^3J_{3,4H_E}$ = 16.8 Hz, 1 H, 4-H$_E$), 6.44 (ddd, $^3J_{2,3}$ = 10.9, $^3J_{3,4H_E}$ = 10.1, $^3J_{3,4H_Z}$ = 16.8 Hz, 1 H, 3-H), 6.72 ("d", $3J_{2,3}$ = 11 Hz, 1 H, 2-H), 7.12–7.30 (m, 10 H of 2 C$_6$H$_5$).

^{13}C NMR (62.9 MHz, CDCl$_3$): δ = 118.6 (d, C-2), 127.3 (d), 127.4 (d), 127.5 (d), 128.1 (d), 128.2 (d), 128.5 (d), 130.4 (d), 134.9 (d), 139.6 (s, *p*-C of C$_6$H$_5$), 142.0 (s, *p*-C of C$_6$H$_5$), 143.1 (s, C-1).

Experiment 39 (AB 119)
Benzyl-fluoren-9-ylidene-amine (36)

36

In analogy with a literature procedure,[109] fluorenone (1.00 g, 7.3 mmol) and benzylamine (2.92 g, 27.3 mmol) were dissolved in dichloromethane (40 mL), and TiCl$_4$ (5.5 mL of an ca. 0.82 M dichloromethane solution, ca. 4.6 mmol) was added dropwise under nitrogen, keeping the temperature at 10–20 °C. The resulting suspension was stirred for an additional 30 min and filtered through a pad of magnesium oxide/silica gel (1:4) (10 g). The filtrate was concentrated *in vacuo* (18 mbar)and recrystallized from diethyl ether/petroleum ether giving 1.19 g (61 %) of the product **36** as yellow needles (m. p. 75–77 °C).

C$_{20}$H$_{15}$N	calcd	C 89.19	H 5.61	N 5.20
(269.1)	found	C 89.15	H 5.63	N 5.20

IR (neat): ν = 3033 (w), 1636 (m), 1597 (m, C=N), 1446 (m), 1299 (m), 788 (m), 746 (s), 721 (vs), 699 (vs), 647 (s), 613 (m) cm^{-1}.

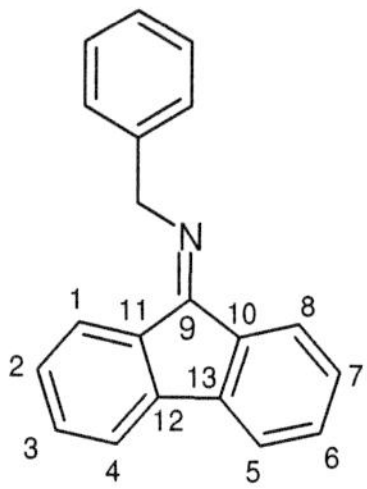

^{1}H NMR (300.1 MHz, CDCl$_3$): δ = 5.32 („s", 2 H, CH$_2$C$_6$H$_5$), 7.18–7.41 (m, 7 H), 7.47–7.54 (m, 3 H), 7.57–7.61 ("d", J = 7.3 Hz, 1 H), 7.82–7.88 (m, 2 H).

^{13}C NMR (75.5 MHz, CDCl$_3$): δ = 56.8 (t, CH$_2$C$_6$H$_5$), 119.4 (d), 120.4 (d, C-4, C-5), 122.8 (d, C-1, C-8), 126.9 (d, p-C of C$_6$H$_5$), 127.7 (d), 128.0 (d), 128.4 (d), 128.5 (d), 131.0 (d), 131.5 (d), 131.9 (s), 138.3 (s), 140.1 (s), 141.1 (s), 143.8 (s, C-12, C-13), 164.0 (s, C=N).

Experiment 40 (AB 120)

(9-Allyl-9H-fluoren-9-yl)-benzylamine (37)

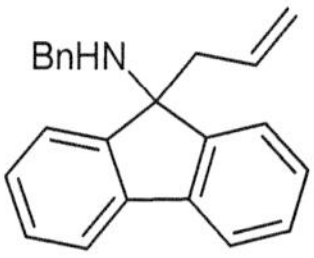

37

The experiment was carried out according to TP 6:.

Imine **36** 0.50 g, 1.86 mmol
Abs. ether 10 mL
Allylmagnesium bromide ca. 4.2 mmol, 2.0 mL of ca. 2.1 M ethereal solution

After the usual work-up 0.59 g ("102 %") of the crude product was isolated as a colourless oil. Flash chromatography on SiO$_2$ (20 mL) (eluting with ethyl acetate/petroleum ether/Et$_3$N 10:90:2, v/v/v) gave 0.50 g (86 %) of a colourless, analytically almost pure oil of the amine **37**.

C$_{23}$H$_{21}$N	calcd	C 88.71	H 6.80	N 4.50
(311.2)	found	C 87.91	H 6.79	N 4.48

^{1}H NMR (300.1 MHz, CDCl$_3$): δ =1.96 (bs, 1 H, NH), 2.69 (ddd, $^3J_{1,2}$ = 7.3, $^4J_{1,3E(Z)}$ = 1.0, $^4J_{1,3E(Z)}$ = 1.3, Hz, 2 H, 1-H), 4.89 ("dd", $^2J_{3E,Z}$ = 2.2, $^3J_{2,3E}$ = 10.1 Hz, 1 H, 3-H$_Z$), 4.93 ("dd", $^2J_{3E,3Z}$ = 2.2, $^3J_{2,3Z}$ = 17.0 Hz, 1 H, 3-H$_E$), 5.51 ("ddt", $^3J_{1,2}$ = 7.4, $^3J_{2,3E}$ = 10.1, $^3J_{2,3Z}$ = 17.4 Hz, 1 H, H-2), 7.13–7.25 (m, 5 H, C$_6$H$_5$), 7.29–7.39 (m, 4 H, 2'-H, 3'-H, 6'-H, 7'-H), 7.49 ("d", J = 7.6 Hz, 2 H, 1'-H, 8'-H), 7.68 ("d", J = 7.3 Hz, 2 H, 4'-H, 5'-H).

^{13}C NMR (75.5 MHz, CDCl$_3$): δ = 45.6 (t, CH$_2$C$_6$H$_5$), 47.4 (t, C-1), 69.9 (s, C-9'), 118.6 (t, C-3), 119.9 (d), 123.6 (d), 126.7 (d, p-C of C$_6$H$_5$), 127.4 (d), 128.0 (d), 128.1 (d), 128.2 (d), 133.0 (d, C-2), 140.5 and 140.9 (2 s, C-10', C-11'), 148.1 (s, C-12', C-13').

Experiment 41 (AB 147, AB 151)
9-Allyl-9-aminofluorene (39)

39

Lithium hexamethyldisilazide for this experiment was prepared as follows:[111]
To a solution of freshly distilled 1,1,1,3,3,3-hexamethyldisilazane (1.84 mmol, 0.296 g) in absolute THF (4 mL) n-butyllithium (1.67 mmol, 1.05 mL of a ca. 1.59 M hexane solution), was added by means of a syringe at 0 °C. The mixture was stirred at room temperature for 40 min and used directly in the next step.

Following the literature procedure,[cf.88,113] to the THF solution of lithium hexamethyldisilazide (ca. 1.67 mmol) fluorenone (300 mg, 1.67 mmol) was under nitrogen added dropwise as a solution in absolute THF (7 mL) within 10 min at 0 °C. The resulting colourless solution was stirred for 2 h at ambient temperature and became yellow. It was then cooled to 0 °C again and allylmagnesium bromide (ca. 2.0 mmol, 2.2 mL of ca. 0.9 M solution) was added dropwise. After stirring overnight (ca. 17 h) at ambient temperature the reaction was quenched by saturated aqueous NaHCO$_3$ solution, the aqueous layer was separated and extracted with diethyl ether (3 × 10 mL). The combined organic layers were washed with brine, dried, and concentrated *in vacuo* (18 mbar) to give 430 mg ("115 %") of a yellow oil. This was purified by column chromatography on SiO_2 (40 g, 12 cm × 3 cm column), eluting with a methanol/dichloromethane/Et_3N mixture (97:3:2, v/v/v). From this, 340 mg (92 %) of an orange oil was isolated. Since this was not spectroscopically pure, the purification was repeated (50 g SiO_2, 14 cm × 3.5 cm column), eluting with ethyl acetate/petroleum ether/Et_3N (30:70:2, v/v/v). The product was isolated as a yellow, spectroscopically pure oil, 90 mg (24 %).

$C_{16}H_{15}N$	calcd	C 86.84	H 6.83	N 6.33
(221.1)	found	C 85.30	H 7.09	N 5.98

MS EI (m/z) calc. for $C_{16}H_{15}N$ 221.1; found 221.1

^{1}H NMR (300.1 MHz, $CDCl_3$): δ = 1.78 (bs, 2 H, NH_2), 2.68 (ddd, $^4J_{2,4HE}$ = 0.9, $^4J_{2,4HZ}$ = 1.4, $^3J_{2,3}$ = 7.3 Hz, 2 H, 2-H), 4.95 ("ddt", $^4J_{2,4HE}$ = 0.9, $^4J_{4HE,4HZ}$ = 2.1, $^3J_{3,4HE}$ = 17.1 Hz, 1 H, 4-H_E), 4.95 ("ddt", $^4J_{2,4HZ}$ = 1.3, $^4J_{4HE,4HZ}$ = 2.1, $^3J_{3,4HZ}$ = 10.1 Hz, 1 H, 4-H_Z), 5.51 ("ddt", $^3J_{2,3}$ = 7.3, $^3J_{3,4HE}$ = 16.6, $^3J_{3,4HZ}$ = 10.1 Hz, 1 H, 3-H), 7.28–7.38 (m, 4 H, 2'-H, 3'-H, 6'-H, 7'-H), 7.48–7.51 (m, 2 H, 1'-H, 8'-H), 7.64–7.67 (m, 2 H, 4'-, 5'-H).

^{13}C NMR (75.5 MHz, $CDCl_3$): δ = 45.5 (t, C-2), 64.8 (s, C-9'), 118.8 (t, C-4), 120.0 (d, C-4', C-5'), 123.4 (d, C-1', C-8'), 127.7 (d, C-3', C-6'), 128.1 (d, C-2', C-7'), 133.3 (d, C-3), 139.3 (s, C-10', C-11'), 150.8 (s, C-12', C-13').

This experiment was repeated (AB 151) with commercially available ca. 1 M THF solution of LiHMDS (Fluka). After usual treatment, a colourless spectroscopically pure oil of the amine **38** was obtained with a yield of 60 %.

Experiment 42 (AB 216)

tert-Butyl (9-allyl-9H-fluoren-9-yl)-carbamate (40)

40

*Typical procedure for the preparation of the fluorenone N-trimethylsilylimine (**38**) (TP 7)[88,113]*

Fluorenone (5.00 g, 27.8 mmol) dissolved in absolute THF (30 mL), and lithium hexamethyldisilazide (ca. 29.2 mmol, 29.2 mL of ca. 1 M THF solution, Fluka) was added to this dropwise within 20 min while cooling with an ice bath. After stirring overnight (20 h) the resulting solution of the imine **38** was used in the following step without purification.

The solution of the imine **38** was cooled to 0 °C and allylmagnesium bromide (ca. 33 mmol, 16.5 mL of ca. 2 M THF solution, Aldrich) was added dropwise. After 18 h stirring at ambient temperature, the reaction was quenched by addition of saturated aqueous $KHCO_3$ solution (60 mL), the aqueous layer was separated and extracted with diethyl ether (4 ×, 350 mL). The combined organic layers were washed with brine, dried over Na_2SO_4 and concentrated *in vacuo* (15 mbar) to give a brown oil (6.42 g). This was dissolved in acetonitrile (40 mL), and, according to a literature procedure,[114] treated with di-*tert*-butyl dicarbonate (6.06 g, 27.8 mmol) and stirred for 17 h. The volatiles were evaporated and the crude product was purified by column chromatography [250 g SiO_2, eluting with ethyl acetate/petroleum ether/Et_3N (from 5:95:2 to 15:85:2, v/v/v)]. The yellow solid isolated (6.50 g) was recrystallized from diethyl ether/petroleum ether to give the title compound as a colourless, spectroscopically and analytically pure solid (5.46 g,61 %), m. p. 119–120 °C.

IR (neat): ν = 3249 (w, NH), 1697 (s, C=O), 1377 (s), 1364 (s), 1164 (s), 734 (s).

$C_{21}H_{23}NO_2$	calcd	C 78.47	H 7.21	N 4.36
(321.4)	found	C 78.33	H 7.07	N 4.39

^{1}H NMR (500 MHz, CDCl$_3$): δ = 0.80 and 1.28 [2 bs, together 9 H, C(CH$_3$)$_3$], 2.56 (bs, 2 H, 2-H), 5.07 (m, 2 H, 4-H), 5.27 (bs, 1 H, NH), 5.61 (bs, 1 H, 3-H), 7.28 ("dt", $^4J_{1',3'}$ = 1.1, $^3J_{2',3'}$ = 7.5 Hz, 2 H, 3'-H, 6'-H), 7.34 ("dt", $^4J_{2',4'}$ = 1.1, $^3J_{2',3'}$ = 7.5 Hz, 2 H, 2'-H, 7'-H), 7.48 (bm, 2 H, 1'-H, 8'-H), 7.64 ("d", $^3J_{3',4'}$ = 7.5 Hz, 2 H, 4'-H, 5'-H).

^{13}C NMR (75.5 MHz,CDCl$_3$): δ = 27.8 [bq, C(CH$_3$)$_3$], 45.1 [bs, C(CH$_3$)$_3$], 65.2 (s, C-1), 79.4 (t, C-2), 119.8 (t, C-4), 123.1 (bd), 127.4 (d), 128.1 (d), 132.3 (bd, C-3), 139.7 (s, C-10', C-11'), 148.7 (bs, C-12', C-13'), 154.6 (bs, C=O).

Some signals appeared broad due to the presence of rotamers. Measurements of NMR spectra in deuterated dimethylsulfoxide did not improve the quality of the spectra significantly:

^{1}H NMR (500 MHz, DMSO, 430 K): δ = 0.68 and 1.23 [2 bs, together 9 H, C(CH$_3$)$_3$], 2.71 (d, J = 5.9 Hz, 2 H, 2-H), 4.72 (bs, 1 H, 4-H$_E$), 4.74 (m, 1 H, 4-H$_z$), 5.12 (m, 1 H, 3-H), 7.26-7.34 (m, 4 H, 2'-H, 3'-H, 6'-H, 7'-H), 7.39-7.45 (m, 2 H, 1'-H, 8'-H), 7.73 („d", $^3J_{3',4'}$ = 7.4 Hz, 2 H, 4'-H, 5'-H).

^{13}C NMR (75.5 MHz,DMSO): δ = 27.9 [bq, C(CH$_3$)$_3$], 43.6 [s, C(CH$_3$)$_3$], 65.3 (s, C-1), 77.6 (t, C-2), 118.5 (d), 119.7 (t, C-4), 122.5 (d), 127.2 (d), 127.5 (d), 132.2 (d, C-3), 139.6 (s, C-10', C-11'), 148.7 (bs, C-12', C-13'), 154.62 (bs, C=O).

Recording of the NMR spectra at 430 K (157 °C) using deuterated dimethylsulfoxide as a solvent resulted in a clear ^{1}H NMR spectrum:

^{1}H NMR (500 MHz, DMSO, 430 °C): δ = 1.04 [s, 9 H, C(CH$_3$)$_3$], 2.77 (d, J = 7.0 Hz, 2 H, H-2), 4.76 (s, 1 H, 4-H$_E$), 4.78 (m, 1 H, 4-H$_z$), 5.26 (m, 1 H, 3-H), 6.28 (bs, 1 H, NH), 7.23 ("t", $^3J_{2',3'}$ = 7.4 Hz, 2 H, 3'-H, 6'-H), 7.29 ("t", $^3J_{2',3'}$ = 7.4 Hz, 2 H, 2'-H, 7'-H), 7.45 („d", $^3J_{1',2'}$ = 7.4 Hz, 2 H, 1'-H, 8'-H), 7.65 („d", $^3J_{3',4'}$ = 7.4 Hz, 2 H, 4'-, 5'-H).

Experiment 43 (AB 179)

tert-Butyl [9-(2,3-dihydroxypropyl)-9H–fluoren-9-yl]-carbamate
(41)

41

According to a know procedure,[99] the suspension of the allyl derivative **40** (3.38 g
,10.5 mmol), potassium hexacyanoferrate (10.36 g, 31.5 mmol) and potassium osmate
dihydrate (25 mg, 0.08 mmol) in *tert*-butanol/water (1:1, 240 mL) was cooled to 0 °C and
potassium carbonate (4.35 g, 31.5 mmol) was added. The reaction was monitored by TLC
(ethyl acetate/petroleum ether 20:80). After 5 days at ambient temperature, sodium sulfite
was added (7.0 g), and the suspension formed was stirred for an additional hour. The
mixture was then extracted with ethyl acetate (3 × 100 mL). The combined organic layers
were dried over Na_2SO_4. The solvent was removed *in vacuo* (15 mbar) to give 3.62 g (97 %)
of a colourless solid that was filtered through a SiO_2 column (eluting with ethyl
acetate/petroleum ether 50:50, v/v). The product **41** was isolated as a colourless,
specroscopically and analytically almost pure foam (3.21 g, 86 %), m. p. 51–54 °C.

IR (neat): ν = 3337 (b, OH), 1674 (vs, C=O), 1390 (s), 1364 (s), 1164 (s), 1051 (s), 1028 (s),
734 (vs) cm^{-1}.

$C_{21}H_{25}NO_4$	calcd	C 70.96	H 7.09	N 3.94
(355.4)	found	C 70.00	H 7.15	N 3.27

^{1}H NMR (500 MHz, CDCl$_3$) : δ = 0.90 [bs, 9 H, C(CH$_3$)$_3$], 2.30 (m, 2 H, 1-H), 2.32 ("dd", J =
10.2, J = 14.8 Hz, 2 H, 1-H), 2.82 (bs, 1 H, NH), 3.30 (m, 2 H, 3-H), 3.35 (bs, 2 H), 7.27-7.47
(m, 5 H), 7.61-7.69 (m, 3 H).

Some peaks in the ^{1}H NMR spectrum were very broad. Measurements at 400 K (127 °C)
gave a clear spectrum:

^{1}H NMR (500 MHz, $C_2D_2Cl_4$, 400 K) : δ = 1.05 [s, 9 H, C(CH$_3$)$_3$], 1.72 (bs, 1 H, OH), 1.89 (dd, $^2J_{1a,1b}$ = 14.5, $^3J_{1a,2}$ = 2.5 Hz, 1 H, 1-H$_a$), 2.20 (bs, 1 H, OH), 2.36 (dd, $^2J_{1a,1Hb}$ = 14.5, $^3J_{1b,2}$ = 8.7 Hz, 1 H, 1-H$_b$), 3.21 (dd, $^3J_{3a,2}$ = 6.8, $^2J_{3a,3b}$ = 11.0 Hz, 1 H, 3-H$_a$), 3.27 (dd, $^3J_{3b,2}$ = 3.9, $^2J_{3a,3b}$ = 11.0 Hz, 1 H, 3-H$_b$), 3.66 (m, 1 H, 2-H), 7.25 ("dd", J = 7.2, J = 7.8 Hz, 2 H), 7.32-7.43 (m, 2 H), 7.47 ("d", J = 7.3 Hz, 1 H), 7.58-7.63 (m, 3 H).

^{13}C NMR (125 MHz,CDCl$_3$): δ = 27.7 [q , C(CH$_3$)$_3$], 42.3 (t, C-1) , 60.4 (s, C-9'), 65.8 (d, C-2) 67.2 [s, C(CH$_3$)$_3$], 70.1 (t, C-3), 119.9 (d), 120.1 (d), 122.3 (d), 123.7 (d), 127.3 (d), 127.7 (d), 127.9 (d), 128.2 (d), 138.8 and 140.0 (2 s, C-10', C-11'), 148.3 (s, C-12', C-13'), 155.5 (s, C=O).

Experiment 44 (AB 181)
(9-*tert*–Butoxycarbonylamino)-(9H-fluoren-9-yl)-acetic acid (42)

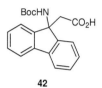

42

The reaction was performed according to TP 5:

tert-Butyl [9-(2,3-dihydroxy-propyl)-9H–fluoren-9-yl]-carbamate (**41**)
1.54 g, 4.35 mmol
THF/water mixture 60:40, 20 mL
NaIO$_4$ 1.40 g , 6.52 mmol
tert-Butanol 10 mL
2-Methyl-2-buten 10 mL
NaClO$_2$ 0.49 g, 5.44 mmol and 0.25 g, 2.7 mmol
KH$_2$PO$_4$ 0.75 g, 5.44 mmol and 0.38 g, 2.7 mmol

After a work-up, the product was obtained as a colourless solid (1.17 g, 79 %), m. p. 188–192 °C.

IR (neat): ν = 3333 (b), 3206 (b), 3071 (b), 1714 (vs, NC=O), 1654 (vs, C=O), 1449 (s), 1395 (vs), 1367 (s), 1159 (vs), 731 (vs), 653 (s) cm^{-1}.

C$_{20}$H$_{21}$NO$_4$	calcd	C 70.78	H 6.24	N 4.13
(339.4)	found	C 70.79	H 6.20	N 4.03

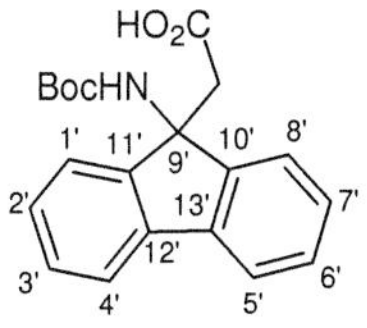

^{1}H NMR (250 MHz, CF$_3$COOD): δ = 1.66 [s, 9 H, (CH$_3$)$_3$], 3.45 (s, 2 H, CH$_2$C=O), 7.47 ("t", J = 7.6 Hz, 2 H), 7.62 ("t", J = 7.6 Hz, 2 H), 7.78 ("d", J = 7.6 Hz, 2 H), 7.84 ("t", J = 7.7 Hz, 2 H).

^{13}C NMR (62.9 MHz, CF$_3$COOD): δ = 28.5 [q, C(CH$_3$)$_3$], 41.3 (t, CH$_2$C=O), 92.8 [s, C(CH$_3$)$_3$], 66.0 (s, C-9'), 123.5 (d, C-4', C-5'), 127.4 (d, C-1', C-8'), 131.4 (d, C-3', C-6'), 134.2 (d, C-2', C-7'), 141.9 (s, C-10', C-11'), 142.0 (s, C-12', C-13'), 142.4 (s, NCOO^{t}Bu), 179.3 (s, COOH).

Experiment 45 (AB 103)
N-Isopropylidene-(4-methoxybenzyl)-amine (43)

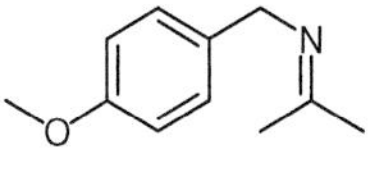

43

According to the literature,[154] a round-bottomed flask was charged with p-methoxybenzylamine (0.50 g, 3.6 mmol) and acetone (6 mL) as well as Na$_2$SO$_4$ (ca. 1 g). The resulting suspension was stirred overnight (16 h) at room temperature. Then this suspension was filtered and the solvent was removed in vacuo (18 mbar) to give 0.63 g (quant.) of the imine **43** as a colourless, spectroscopically pure oil that after one week proved to be unstable on storage at 4 °C.

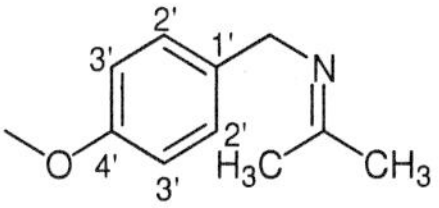

^{1}H NMR (300.1 MHz, CDCl$_3$): δ = 1.92 and 2.07 [2 s, 3 H each, N=C(CH$_3$)$_2$], 3.80 (s, 3 H, CH$_3$O), 4.39 ("s", 2 H, CH$_2$C$_6$H$_4$), 6.86–6.89 (m, 2 H, 3'-H), 7.17–7.20 (m, 2 H, 2'-H).

^{13}C NMR (75.5 MHz, CDCl$_3$): δ = 18.7 and 29.5 [2 q, N=C(CH$_3$)$_2$], 54.9 (t, CH$_2$C$_6$H$_4$), 55.3 (q, CH$_3$O), 113.9 (d, C-3'), 128.9 (d, C-2'), 132.6 (s, C-1'), 158.4 (s, C-4'), 167.8 [s, N=C(CH$_3$)$_2$].

Experiment 46 (AB 106)

2-(4-Methoxybenzylamino)-2-methyl-4-pentene (44)

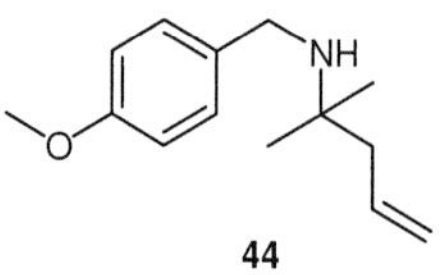

44

The allylation was carried out in accordance with TP 6:

Imine **42** 0.50 g, 2.8 mmol

Abs. diethyl ether 7 mL

Allylmagnesium bromide ca. 5.7 mmol, 2.7 mL of ca. 2.1 M ethereal solution

The usual work-up afforded 0.45 g (72 %) of a yellow semisolid. Flash chromatography on silica gel (3 cm column, eluting with a mixture methanol/dichloromethane 10:90) gave 0.29 g (47 %) of a colourless, analytically pure semisolid of the allyl derivative **44**.

$C_{14}H_{21}NO$	calcd	C 76.67	H 9.65	N 6.39
(219.2)	found	C 76.53	H 9.78	N 6.14

IR (neat): ν = 2960 (m, NH), 1612 (m), 1510 (vs), 1440 (m), 1300 (m), 1243 (vs), 1172 (s), 1035 (s, C-O), 1000 (m), 915 (m), 822 (s), 747 (m), 697 (m), 625 (s), 557 (s) cm^{-1}.

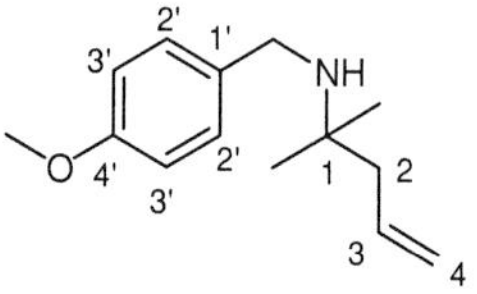

^{1}H NMR (300.1 MHz, CDCl$_3$): δ = 1.13 [s, 6 H, (CH$_3$)$_2$], 2.23 ("d", $^3J_{2,3}$ = 6 Hz, 2 H, 2-H), 3.64 ("s", 2 H, CH$_2$C$_6$H$_4$), 3.76 (s, 3 H, OCH$_3$), 5.08 (m, 2 H, 4-H), 5.86 (m, 1 H, 3-H), 6.83 ("d", $^3J_{2,3}$ = 9 Hz, 2 H, H-3'), 7.25 ("d", $^3J_{2,3}$ = 9 Hz, 2 H, H-2').

^{13}C NMR (75.5 MHz, CDCl$_3$): δ = 27.1 [q, (CH$_3$)$_2$], 45.2 (t, C-2), 46.0 (s, C-1), 52.8 (t, CH$_2$C$_6$H$_4$), 55.2 (q, CH$_3$O), 113.8 (t, C-3'), 117.7 (t, C-4), 129.4 (t, C-2'), 133.3 (s, C-1'), 134.8 (d, C-3), 158.5 (s, C-4').

7.5   Experiments concerning Chapter 4

Spiro[azetidine-2,9'-9*H*-fluorene]-4-one (46)

46

Experiment 47 (AB 194)

A solution of the *N*-trimethylsilylimine **38** was prepared according to TP 7:

Fluorenone 1.00 g, 5.56 mmol

Abs. THF 15 mL

LiHMDS ca. 6.1 mmol, 6.1 mL of ca. 1.0 M THF solution

*Typical procedure for the preparation of THF solution of the lithium enolate of ethyl acetate (**45**) (TP 8):[121]*

To diisopropylamine (0.84 g, 1.2 mL, 8.33 mmol) dissolved in absolute THF (16 mL) *n*-butyllithium (ca. 8.3 mmol, 5.3 mL of ca. 1.58 M in hexane, Merck) was added at 0 °C under nitrogen. The resulting solution was stirred for 15 min at this temperature, then cooled to –78 °C and treated with ethyl acetate (0.90 mL, ca. 9.2 mmol) followed by stirring for 1 h at –78 °C.

A solution of the imine **38** was added dropwise to this solution of the lithium enolate **45** at –78 °C, then the mixture was allowed to warm up slowly to room temperature. The mixture was left with stirring for 16 h (it became dark-green) and then treated with saturated aqueous NaHCO₃ solution (20 mL). The aqueous layer was extracted with ethyl acetate (3 × 20 mL). Removal of the organic solvents (18 mbar) gave 1.53 g of a brown oil which was purified by column chromatography (SiO₂, 25 g, 11 cm × 3 cm), eluting with ethyl acetate/petroleum ether/Et₃N (10:90:2, v/v/v). Fluorenone (0.32 g, 32 %) was isolated as a yellow solid, as well as a yellow oil of a mixed fraction (0.69 g). This oil was purified once again on silica gel (25 g, column 17 cm × 2.5 cm) and gave a yellowish solid (0.200 g). Recrystallisation from ethyl acetate/petroleum ether gave 0.14 g (11 %) of the β-lactam **46** as a colourless, spectroscopically and analytically pure solid, m. p. 183–185 °C.

$C_{15}H_{11}NO$	calcd	C 81.43	H 5.01	N 6.33
(221.3)	found	C 81.48	H 5.01	N 6.31

IR (neat): ν = 3157 (bw, NH), 1740 (s, C=O, amide I), 1713 (vs), 1450 (m), 1407 (m), 1272 (m), 1233 (m), 984 (m), 747 (vs), 722 (vs) cm^{-1}.

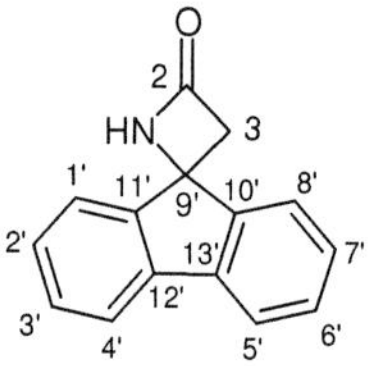

^{1}H NMR (300.1 MHz, CDCl$_3$): δ = 3.42 (s, 2 H, 3-H), 6.21 (bs, 1 H, NH), 7.34 ("t", 2 H, $^3J_{2',3'}$ = 7.3 Hz, 3'-H, 6'-H), 7.40 ("t", $^3J_{2',3'}$ = 7.3 Hz, 2 H, 2'-H, 7'-H), 7.59 ("d", $^3J_{1',2'}$ = 7.5 Hz, 2 H, 1'-H, 8'-H), 7.64 ("d", $^3J_{3',4'}$ = 7.5 Hz, 2 H, 4'-H, 5'-H).

^{13}C NMR (75.5 MHz, CDCl$_3$): δ = 51.1 (t, C-3), 61.0 (s, C-9'), 120.3 (d, C-4', C-5'), 123.0 (d, C-1', C-8'), 128.2 (d, C-3', C-6'), 129.5 (d, C-2', C-7'), 139.3 (s, C-10', C-11'), 144.9 (s, C-12', C-13'), 167.7 (s, C-2).

Experiment 48 (AB 221)

A solution of the imine **38** was prepared according to TP 7:

Fluorenone 1.00 g, 5.56 mmol
LiHMDS ca. 6.1 mmol, 6.1 mL of ca. 1.0 M THF solution

A solution of the lithium enolate of ethyl acetate (**45**) was prepared according to TP 8:

Diisopropylamine 1.85 g, 2.6 mL, 18.4 mmol
n-Butyllithium 11.6 mL of ca. 1.58 M hexane solution, ca. 18.4 mmol
Ethyl acetate 1.47 g, ca. 16.7 mmol

The solution of the enolate **45** was added dropwise to the solution of the imine **38** at 0 °C, and the mixture was stirred for 20 h at ambient temperature. The reaction was quenched with saturated aqueous NH$_4$Cl solution (10 mL), the aqueous layer was separated and extracted with ethyl acetate (3 × 15 mL). Removal of the organic solvents *in vacuo* (18 mbar) gave 1.94 g of a brown oil which was purified by column chromatography on silica gel (95 g, column 27 cm × 3.5 cm), eluting with of ethyl acetate/petroleum ether/Et₃N (10:90:2, v/v/v). The first fraction, a yellow oil (0.225 g) was ethyl acetoacetate. The second fraction gave 0.51 g of a yellowish solid, which in recrystallization from ethyl acetate/petroleum ether gave 0.49 g (2.2 mmol, 40 %) of the β-lactam **46** as an analytically pure, colourless solid, m. p. 180–183 °C.

$C_{15}H_{11}NO$	calcd	C 81.43	H 5.01	N 6.33
(221.3)	found	C 81.33	H 5.11	N 6.34

IR and NMR data were in ccordance with those given in the Exp. 47.

3-(9-amino-9-fluorenyl)-azetidin-4-one-2-spiro-9'-fluorene (47)

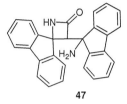

47

Experiment 49 (AB 225)

A solution of the imine **38** was prepared according to TP 7:

Fluorenone 0.50 g, 2.8 mmol
LiHMDS ca. 3.1 mmol, 3.1 mL of ca. 1.0 M THF solution

A solution of the lithium enolate of ethyl acetate (**45**) was prepared according to TP 8:

Diisopropylamine 0.84 g, 1.2 mL, 8.3 mmol
n-Butyllithium ca. 8.3 mmol, 5.3 mL of ca. 1.58 M hexane solution
Ethyl acetate 0.50 g, ca. 5.6 mmol

The solution of the lithium enolate **45** was added dropwise to the solution of the imine **38** at 0 °C. The mixture was left with stirring for 17 h at room temperature and changed the colour from orange to red-brown. The reaction was quenched by ice-cold water (5 mL), the aqueous layer was separated and extracted with ethyl acetate (4 ×, 50 mL). Removal of the organic solvents gave 0.80 g of a brown oil which was purified by column chromatography on SiO₂ (24 g), eluting with ethyl acetate/petroleum ether/Et₃N (20:80:2 to 50:50:2, v/v/v).

The first fraction, a yellowish solid (0.212 g, 0.96 mmol, 34 %), was the β-lactam **46** (AB 225-2) (NMR data were in accordance with those from the experiment AB 221).

The second fraction, a yellowish solid (0.140 g, 0.35 mmol, 25 %), m. p. 210–213 °C, consisted of the analytically almost pure 3,4-disubstituted azetidinone **47** (AB 225-3). Its structure was established based on X-ray crystal structure analysis.

$C_{28}H_{20}N_2O$	calcd	C 83.98	H 5.03	N 7.00
(400.5)	found	C 83.17	H 5.03	N 6.88

IR (neat): ν = 3147 (bw, NH), 3012 (bm, NH$_2$), 1736 (vs, C=O), 1448 (m), 1277 (m), 1233 (m), 928 (m), 727 (vs) cm^{-1}.

MS (ESI, m/z): calc. for $C_{28}H_{20}N_2O$ 400.16; found 401.16 [M+H]$^+$

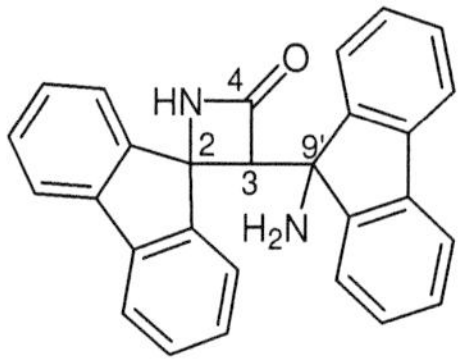

^{1}H NMR (300 MHz, MeOD): δ = 4.71 (s, 1 H, 3-H), 5.73 ("d", J = 7.6 Hz, 1 H), 6.44 ("t", J = 7.5 Hz, 1 H), 6.61–6.70 (m, 2 H), 6.93 ("t", J = 7.5 Hz, 2 H), 7.06 ("dd", J = 3.4, J = 7.5 Hz, 2 H), 7.32–7.47 (m, 6 H,), 7.64–7.68 (m, 1 H), 7.94–7.96 (m, 1 H).

^{13}C NMR (75 MHz, MeOD): δ = 63.9 (s, C-2 or C-9'), 66.5 (s, C-2 or C-9'), 70.1 (d, C-3); 119.7, 120.3, 120.5, 121.9, 123.1, 123.5, 125.5, 125.7, 127.0, 128.1, 128.6, 128.9, 129.0, 129.1, 129.9, 130.0 (16 d); 140.5, 141.1, 141.6, 141.6, 143.0, 146.8, 147.6, 148.8 (8 s); 171.8 (s, C-4).

Experiment 50 (AB 235)

A solution of the *N*-trimethylsilylimine **38** was prepared according to TP 7:

Fluorenone 1.00 g, 5.56 mmol
LiHMDS ca. 6.1 mmol, 6.1 mL of ca. 1.0 M THF solution

A solution of the lithium enolate of ethyl acetate (**45**) was prepared according to TP 8:

Diisopropylamine 1.85 g, 2.6 mL, 18.4 mmol
n-Butyllithium 11.6 mL of ca. 1.58 M hexane solution, ca. 18.0 mmol
Ethyl acetate 1.47 g, ca. 16.7 mmol

A solution of the imine **38** was added dropwise to the solution of the lithium enolate of ethyl acetate **45** at –60 °C, the mixture was left with stirring for 17 h at 0 °C during which time the colour changed from orange to brown. The reaction was quenched by ice-cold saturated aqueous NH_4Cl solution (15 mL), the aqueous layer was separated and extracted with ethyl acetate (4 ×, 60 mL). Removal of the organic solvents *in vacuo* (0.1 mbar) gave 1.41 g of a green-brown sticky oil, which after precipitation with diethyl ether gave a colourless solid, the product **47**, 0.170 g (0.42 mmol, 15 %), m. p. 210–213 °C (AB 235-2). Additional product was isolated by repeated treatment of the concentrated mother liquid with diethyl ether (AB 235-3); 0.429 g (1.07 mmol, 38 %), m. p. 209–214 °C. The total yield was 0.600 g (53 %); analytical data of the combined samples are given.

$C_{28}H_{20}N_2O$	calcd	C 83.98	H 5.03	N 7.00
(400.5)	found	C 83.55	H 5.06	N 6.93

IR (neat): ν = 3328 (bw, NH), 3038 (bm, NH_2), 1734 (vs, C=O, amide I), 1448 (m), 1279 (m), 1207 (m), 929 (m), 732 (vs) cm^{-1}.

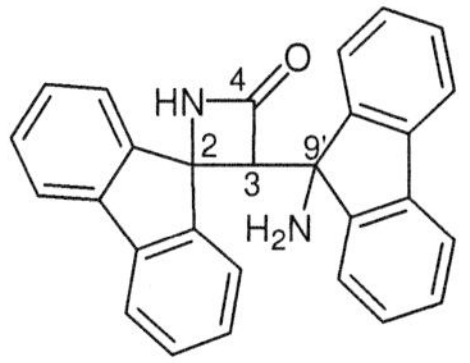

^{1}H NMR (250 MHz, DMSO-d_6): δ = 2.27 (s, 2 H, NH_2), 4.63 (s, 1 H, 3-H), 5.69 (d, *J* = 7.5 Hz, 1 H), 6.45 ("dd", *J* = 7.1, *J* = 7.5 Hz, 1 H), 6.66 (m, 2 H), 6.92 (m, 2 H), 7.09 ("d", *J* = 7.5 Hz,

1 H), 7.18 ("d", J = 7.5 Hz, 1 H), 7.28–7.49 (m, 6 H), 7.66 (m, 1 H), 7.82 (m, 1 H), 8.46 (s, 1 H, NH).

^{13}C NMR (63 MHz, DMSO-d$_6$): δ = 63.0 (s, C-2 or C-9'), 64.4 (s, C-2 or C-9'), 69.1 (d, C-3); 118.5, 118.8, 119.3, 120.3, 122.1, 122.2, 124.1, 125.4, 126.4, 127.1, 127.3, 127.4, 127.5, 128.0, 128.5 (16 d), 138.2, 139.0, 139.6, 140.9, 141.0, 146.7, 147.4, 148.4 (8 s), 168.7 (s, C-4).

Experiment 51 (AB 264)

3-Chloro-1-(4-methoxyphenyl)-spiro[azetidine-2,9'-9H-fluoren]-4-one (50)

50

Typical procedure for the preparation of 3-chlorosubstituted 2-azetidinones (TP 9): [125]

Fluorenone *N*-(*p*-methoxyphenyl)-imine (**48**) (4.40 g, 16.1 mmol) was dissolved in dry dichloromethane (60 mL) followed by the addition of Et$_3$N (1.95 g, 2.68 mL, 19.3 mmol). The reaction mixture was cooled to –20 °C, and chloroacetyl chloride (1.82 g, 1.28 mL, 16.1 mL) in dry dichloromethane (5 mL) was added dropwise within 45 min. The resulting solution was stirred for 3 h at –20 °C, then cooled to –30 °C and quenched by addition of saturated aqueous NaHCO$_3$ solution (ca. 40 mL). The aqueous layer was separated and extracted with dichloromethane (4 × 20 mL). The combined organic layers were washed with water, dried over Na$_2$SO$_4$, and concentrated *in vacuo* (18 mbar) to give 5.59 g (96 %) of a yellowish solid. This was recrystallized from methanol/petroleum ether to give 4.91 g (84 %) of an analytically pure, colourless solid, m. p. 171–174 °C.

C$_{22}$H$_{16}$ClNO$_2$	calcd	C 73.03	H 4.46	N 3.87	Cl 9.80
(361.8)	found	C 72.92	H 4.54	N 3.84	Cl 9.60

IR (neat): ν = 3010 (w), 1754 (vs, C=O), 1510 (vs, C–O), 1436 (m), 1379 (s), 1248 (vs), 1168 (vs), 829 (vs), 748 (vs), 728 (s) cm^{-1}.

^{1}H NMR (300 MHz, CDCl$_3$): δ = 3.63 (s, 3 H, OCH$_3$) 5.31 (s, 1 H, 3-H), 6.57–6.63 ("d", 2 H, $^3J_{2',3'}$ = 9.1 Hz, 3'-H), 6.87–6.93 ("d", 2 H, $^3J_{2',3'}$ = 9.2 Hz, 2'-H), 7.28–7.36 (m, 2 H, 3'-H, 6'-H), 7.44–7.59 (m, 4 H, 1'-H, 2'-H, 7'-H, 8'-H), 7.77–7.81 (m, 2 H, 4'-, 5'-H).

^{13}C NMR (75.5 MHz, CDCl$_3$): δ = 55.2 (q, OCH$_3$), 65.0 (d, C-3), 71.8 (s, C-2), 114.2 (d, C-3'), 118.9 (d, C-2'), 120.6 (d), 120.8 (d), 122.9 (d), 127.0 (d), 127.6 (d), 128.4 (d), 129.4 (s, C-1'), 130.3 and 130.4 (2 d, C-2', C-7'), 138.1 and 140.3 (2 s, C-10', C-11'), 141.0 and 141.3 (2 s, C-12', C-13'), 156.6 (s, C-4'), 160.6 (s, C-4).

Experiment 52 (AB 276)

1-(4-Methoxyphenyl)-spiro[azetidine-2,9'-9*H*-fluoren]-4-one (51)

51

Typical procedure (TP 10) for the removal of chlorine from C-3 of β-lactams:[126]

3-Chloro-1-(4-methoxyphenyl)-spiro[azetidine-2,9'-[9H]fluoren]-4-one (**50**) (4.63 g, 12.8 mmol) was dissolved in dry benzene (40 mL) followed by the addition of *tris*(trimethylsilyl)silane (Fluka) (4.78 g, 5.9 mL, 19.2 mmol). The solution was brought to reflux, and AIBN (60 mg, 0.38 mmol) was added in one portion. The mixture was heated under reflux for 3 h, then additional 20 mg of AIBN was added and heating under reflux was continued for 1 h. A colourless solid formed after cooling to room temperature. The volatiles were removed *in vacuo* (50 mbar), the remaining yellowish solid was suspended in petroleum ether and then collected on a glass filter (the petroleum ether phase was discarded). The solid was dissolved on the filter with portions of chloroform (4 × 25 mL). The filtrate was concentrated to 50 mL, and the solution was passed through a silica gel column (3 cm). Silica gel was washed with chloroform again (50 mL), and the filtrate was concentrated *in vacuo* (100 mbar). The rest was precipitated with petroleum ether and gave 3.61 g (86 %) of the product as a colourless, spectroscopically pure solid, m. p. 231–232 °C.

$C_{22}H_{17}NO_2$ calcd   C 80.71   H 5.23   N 4.28

(327.4)   found   C 79.50   H 5.26   N 4.11

IR (neat): ν = 3046 (bw), 2835 (w), 1738 (vs, C=O), 1510 (vs), 1440 (m), 1365 (vs), 1298 (s), 1247 (vs), 827 (s), 750 (s), 729 (s) cm^{-1}.

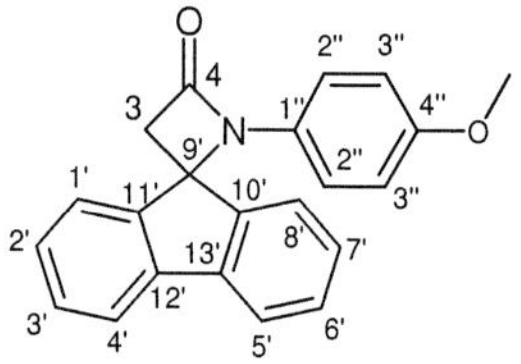

^{1}H NMR (500 MHz, CDCl$_3$): δ = 3.54 (s, 2 H, 3-H), 3.63 (s, 3 H, OCH$_3$), 6.60 ("d", 2 H, $^3J_{2'',3''}$ = 9.2 Hz, 3''-H), 6.86 ("d", 2 H, $^3J_{2'',3''}$ = 9.2 Hz, 2''-H), 7.32 (m, 2 H, 3'-H, 6'-H), 7.45 (m, 2 H, 2'-H, 7'-H), 7.53 ("d", $^3J_{1',2'}$ = 8.2 Hz, 2 H, 1'-H, 8'-H), 7.78 ("d", $^3J_{3',4'}$ = 6.2 Hz, 2 H, 4'-, 5'-H).

^{13}C NMR (75.5 MHz, CDCl$_3$): δ = 50.3 (t, C-3), 55.3 (q, CH$_3$O), 65.2 (s, C-9'), 114.1 (d, C-3''), 118.4 (d, C-2''), 120.6 (d, C-4', C-5'), 123.1 (d, C-1', C-8'), 128.4 (d, C-3', C-6'), 129.7 (d, C-2', C-7'), 130.7 (s, C-1''), 140.1 (s, C-10', C-11'), 142.9 (s, C-12', C-13'), 155.9 (s, C-4''), 163.8 (s, C-4).

Experiment 53 (AB 298)

Spiro[azetidine-2,9'-9H-fluoren]-4-one (46)

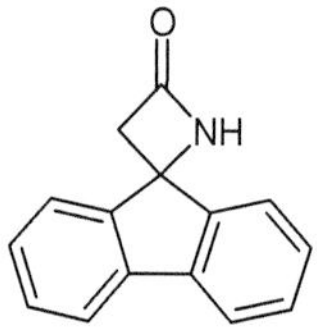

46

Typical procedure (TP 11) for the oxidative cleavage of the N-p-methoxyphenyl protecting group:[101,102]

1-(4-Methoxyphenyl)-spiro[azetidine-2,9'-[9H]fluoren]-4-one (**51**) (500 mg, 1.53 mmol) was dissolved in acetonitrile (40 mL) and cooled to −10 °C. To this, a solution of ceric ammonium nitrate (2.514 g, 4.59 mmol) in water (15 mL) was added dropwise within 15 min, followed by stirring for 1.5 h at −10 °C. Then ethyl acetate (30 mL) as well as water (10 mL) were added, and the aqueous phase was separated followed by extraction with ethyl acetate (20 mL). The combined organic layers were washed subsequently with saturated aqueous solutions of

NaHCO$_3$ (20 mL), NaHSO$_3$ (2 × 20 mL) and NaCl (20 mL). Drying over Na$_2$SO$_4$ and concentration *in vacuo* (15 mbar) gave 460 mg of a brown oil which was purified by column chromatography on SiO$_2$ (15.5 g, 7 cm × 2.5 cm), eluting with ethyl acetate/petroleum ether (20:80 to 50:50). Repeated recrystallization of the main fraction (chloroform/petroleum ether) gave the title compound as a spectroscopically and analytically pure solid (203 mg, 60 %), m. p. 175–180 °C (183–185 °C, Exp. 47; 180-183 °C, Exp. 48).

C$_{15}$H$_{11}$NO	calcd	C 81.43	H 5.01	N 6.33
(221.3)	found	C 81.39	H 5.04	N 6.29

IR, ^{1}H and ^{13}C NMR data were in accordance with those from experiments 47 and 48.

Experiment 54 (AB 290)

3-Chloro-1-(4-methoxyphenyl)-4,4-diphenylazetidin-2-one (52)

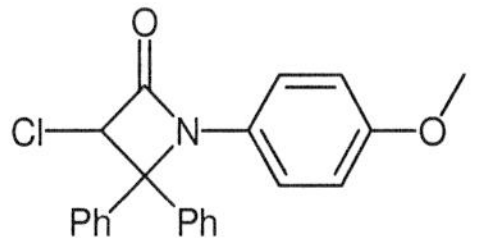

52

The 3-chloro-substituted β-lactam **52** was prepared according to TP 9:

Benzophenone *N*-(*p*-methoxyphenyl)-imine (**49**) 4.99 g, 18.2 mmol

Et$_3$N 2.20 g, 3.02 mL, 21.8 mmol

Chloroacetyl chloride 2.05g, 1.44 mL, 18.2 mmol

The reaction was run for 3 h at –20 °C. The standard work-up gave a crude oily product which was precipitated with petroleum ether to give a greenish solid which on recrystallization from chloroform/petroleum ether yielded 4.89 g (74 %) of a colourless, analytically almost pure solid of **52**, m. p. 143–147 °C.

C$_{22}$H$_{18}$ClNO$_2$	calcd	C 72.62	H 4.99	N 3.85	Cl 9.74
(363.8)	found	C 72.07	H 4.54	N 3.84	Cl 9.60

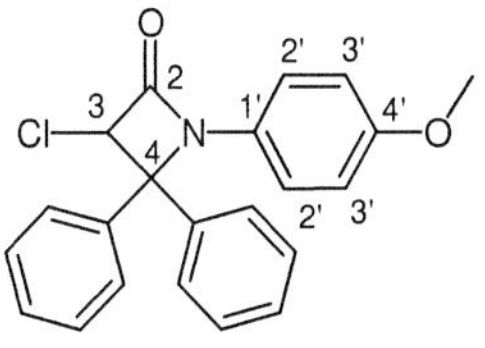

^{1}H NMR (250.1 MHz, CDCl$_3$): δ = 3.72 (s, 3 H, OCH$_3$), 5.39 (s, 2 H, 3-H), 6.73 ("d", 2 H, $^3J_{2',3'}$ = 9.1 Hz, 3'-H), 7.26 ("d", 2 H, $^3J_{2',3'}$ = 9.1 Hz, 2'-H), 7.24–7.48 (m, 10 H, 2 C$_6$H$_5$).

^{13}C NMR (62.9 MHz, CDCl$_3$): δ = 55.4 (q, CH$_3$O), 69.5 (d, C-3), 73.4 (s, C-4), 114.2 (d, C-3'), 120.3 (d, C-2'), 127.6 (d), 128.0 (d), 128.6 (d, p-C of 2 C$_6$H$_5$), 128.9 (d), 129.4 (d), 130.4 (s, C-1'), 134.7 (s, i-C of C$_6$H$_5$), 138.7 (s, i-C of C$_6$H$_5$), 156.6 (s, C-4'), 161.2 (s, C-2).

Experiment 55 (AB 277)

1-(4-Methoxyphenyl)-4,4-diphenylazetidin-2-one (53)

53

The removal of the chlorine atom was performed according to TP 10:

3-Chloro-substituted β-lactam (**52**) 1.77 g, 4.86 mmol

tris(Trimethylsilyl)silane 1.82 g, 2.25 mL, 7.29 mmol

AIBN 24 mg, 0.15 mmol + 10 mg

The reaction was run for 4h. A colourless solid formed after cooling to room temperature. The volatiles were removed *in vacuo* (50 mbar), and the remaining yellowish solid was suspended in petroleum ether. This suspension was filtered over a glass filter; the petroleum ether filtrate was discarded. The colourless solid obtained was then taken up in chloroform (60 mL) and filtered to remove any undissolved particles. The chloroform filtrate was concentrated *in vacuo* (50 mbar) to a volume of ca. 30 mL and treated with petroleum ether to precipitate the title compound **53** (1.52 g, 95 %) as a colourless, analytically pure solid (m. p. 154–156 °C).

C$_{22}$H$_{19}$NO$_2$	calcd	C 80.22	H 5.81	N 4.25
(329.4)	found	C 80.13	H 5.83	N 4.19

IR (neat): ν = 2950 (w), 1977 (w), 1743 (vs, C=O), 1508 (vs, C–O), 1445 (s), 1365 (s), 1243 (vs), 1027 (s), 831 (s), 756 (s), 700 (vs) cm^{-1}.

^{1}H NMR (500 MHz, CDCl$_3$): δ = 3.54 (s, 2 H, 3-H), 3.63 (s, 3 H, OCH$_3$), 6.60 ("d", 2 H, $^3J_{2',3'}$ = 9.2 Hz, 3'-H), 6.86 ("d", 2 H, $^3J_{2',3'}$ = 9.2 Hz, 2'-H), 7.32 (m, 2 H), 7.45 (m, 2 H), 7.53 ("d", J = 8.2 Hz, 2 H), 7.78 ("d", J = 6.2 Hz, 2 H).

^{13}C NMR (75.5 MHz, CDCl$_3$): δ = 50.3 (t, C-3), 55.3 (q, CH$_3$O), 65.2 (s, C-4), 114.1 (d, C-3'), 118.4 (d, C-4'), 120.6 (d), 123.1 (d), 128.4 (d, p-C of 2 C$_6$H$_5$), 129.7 (d), 130.7 (s, C-1'), 140.1 (s, i-C of C$_6$H$_5$), 142.9 (s, i-C of C$_6$H$_5$), 155.9 (s, C-4'), 163.8 (s, C-2).

Experiment 56 (AB 280)
4,4-diphenylazetidin-2-one (54)

54

The removal of the N-p-methoxyphenyl protecting group was done according to TP 11:

1-(p-Methoxy-phenyl)-4,4-diphenylazetidin-2-one (**53**) 329 mg, 1.00 mmol
CAN 1.644 g, 3.00 mmol

The experiment was performed under conditions described in TP 11. The crude product (243 mg) was isolated as an orange solid. It was purified by column chromatography on SiO$_2$ (4.5 g, 6 cm × 1.5 cm), eluting with ethyl acetate/petroleum ether (30:70 to 40:60, v/v). Recrystallization of the main fraction (chloroform/petroleum ether) gave the title compound as a colourless, spectroscopically and analytically pure solid (86 mg, 39 %), m. p. 193–194 °C.

C$_{15}$H$_{13}$NO	calcd	C 80.69	H 5.87	N 6.27
(223.3)	found	C 80.66	H 5.86	N 6.34

IR (neat): ν = 3165 (m, NH), 3086 (w), 1732 (vs, C=O), 1453 (s), 703 (vs) cm^{-1}.

^{1}H NMR (250 MHz, CDCl$_3$): δ = 3.54 (s, 2 H, 3-H), 7.24–7.38 (m, 10 H of 2 C$_6$H$_5$), 7.58 (bs, 1 H, NH).

^{13}C NMR (63 MHz, CDCl$_3$): δ = 53.4 (t, C-3), 61.7 (s, C-4), 126.2 (d, o-C of 2 C$_6$H$_5$), 127.5 (d, p-C of 2 C$_6$H$_5$) , 128.5 (d, m-C of 2 C$_6$H$_5$), 143.1 (s, i-C of 2 C$_6$H$_5$), 168.0 (s, C=O).

Experiment 57 (AB 310)

3-Chloro-spiro[azetidine-2,9'-9H-fluoren]-4-one (55)

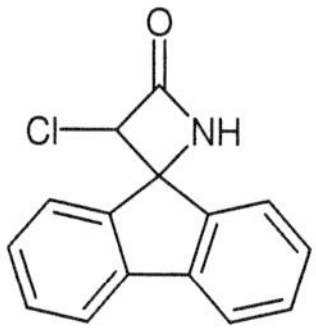

55

A solution of the N-trimethylsilylimine **38** was prepared according to TP 7:

Fluorenone 360 mg, 2 mmol

LiHMDS ca. 2.2 mmol, 2.2 mL of ca. 1.0 M THF solution

The solution of the imine **38** was diluted with dry dichloromethane (3 mL). To this, triethylamine (0.58 mL, 4.2 mmol) was added, the mixture was cooled to –3 °C and treated with chloroacetyl chloride (0.32 mL, 4 mmol) in dry dichloromethane (1 mL). The reaction mixture was left with stirring at 0 °C for 2 h; a colourless solid formed. Saturated aqueous NaHCO$_3$ solution (5 mL) was added as well as 5 mL of water. The aqueous layer was separated and extracted with dichloromethane (3 × 10 mL). The combined organic phases were washed with water (15 mL), dried and concentrated in vacuo (18 mbar) to yield 637 mg of a brown tar. After purification on silica gel (19 g, column 12 cm × 2.3 cm), eluting with ethyl acetate/petroleum ether 20:80 to 25:75 to 100:0 (v/v), 304 mg (59 %) of the title compond **55** as a colorless solid were isolated, m. p. 200–203 °C.

C$_{15}$H$_{10}$ClNO	calcd	C 70.46	H 3.94	N 5.48	Cl 13.87
(255.7)	found	C 70.34	H 4.02	N 5.46	Cl 13.83

IR (neat): ν = 3192 (m), 3072 (w), 2976 (m), 1757 (vs, C=O), 1450 (m), 1363 (m), 1279 (m), 743 (vs), 724 (vs), 590 (vs) cm^{-1}.

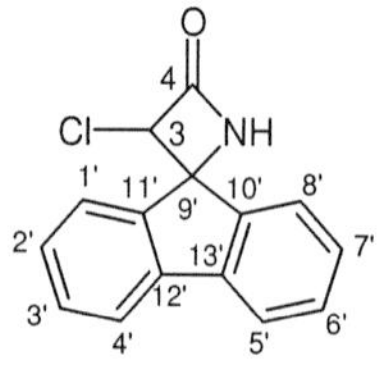

^{1}H NMR (500.1 MHz, DMSO-d$_6$): δ = 5.76 (s, 1 H, 3-H), 7.40–7.53 (m, 4 H of 2 C$_6$H$_4$), 7.58 ("d", J = 7.1 Hz, 1 H), 7.77 ("d", J = 7.3 Hz, 1 H), 7.88 ("d", J = 7.4 Hz, 1 H), 9.22 (bs, 1 H, NH).

^{13}C NMR (125.8 MHz, DMSO-d$_6$): δ = 65.1 (d, C-3), 66.4 (s, C-9'), 120.4 and 120.5 (2 d, C-4', C-5'), 123.8 (d), 126.3 (d), 127.4 (d), 128.3 (d), 129.9 and 130.0 (2 d, C-2', C-7'), 139.5 and 140.4 (2 s, C-10', C-11'), 141.4 and 143.6 (2 s, C-12', C-13'), 163.7 (s, C-4).

7.6   Experiments concerning Chapter 5

2-(9-Fluorenyl)-*N,N,N',N'*-tetramethylsuccindiamide (57)
and
2-(9-amino-9*H*-fluoren-9-yl)-*N,N*-dimethylacetamide (58)

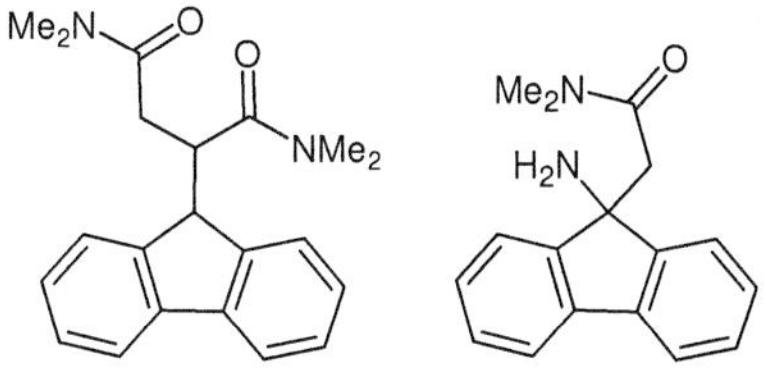

57   58

Experiment 58 (AB 265)

A solution of *N*-trimethylsilylimine **38** was prepared as described in the TP 7:

Fluorenone 180 mg, 1.0 mmol

LiHMDS 1.1 mL of ca. 1.0 M THF solution, ca. 1.1 mmol

Typical procedure for the preparation of the THF solution of the lithium enolate of N,N-dimethylacetamide (56) (TP 12):[132]

To diisopropylamine (152 mg, 0.11 mL, 1.5 mmol), dissolved in absolute THF (4 mL), *n*-butyllithium (ca. 1.5 mmol, 0.95 mL of ca. 1.58 M in hexane, Merck) was added at 0 °C under nitrogen. The resulting solution was stirred for 15 min at 0 °C, *N,N*-dimethylacetamide (131 mg, 0.14 mL, 1.5 mmol) was added followed by stirring for 20 min at 0 °C.

The solution of the imine **38** was added to the solution of the Lithium enolate **56** followed by stirring for 3 h at 0 °C and then 2 h at room temperature. A colourless solid formed. The reaction was quenched by addition of water (10 mL). The aqueous phase was separated and the solid was extracted with ethyl acetate (4 ×, 30 mL). Drying and concentration of the organic layer gave 284 mg of a yellowish oil, which was precipitated by addition of diethyl ether (ca. 5 mL). The solid formed (m. p. 184–186 °C) was recrystallized from chloroform/petroleum ether yielding 90 mg (36 % based on N,N-dimethylacetamide, 27 % based on fluorenone) of the succindiamide derivative **57** as a pale-yellow, analytically almost pure solid, m. p. 179–183 °C.

Diethyl ether from the mother liquid was removed *in vacuo* (15 mbar) yielding 162 mg ("61 %") of the almost pure (^{1}H NMR spectrum) β-amino ester **58** as a yellow oil. The presence of this compound was proven on the basis of the results of the following experiment (Exp. 60).

Analytical and spectroscopical data of the diamide **57**:

IR (neat): ν = 3230 (bw, C(O)N), 2923 (bw), 1634 (vs, C(O)N), 1478 (s), 1445 (s), 1394 (vs), 1132 (vs), 1012 (s), 732 (vs), 517 (vs) cm^{-1}.

$C_{21}H_{24}N_2O_2$	calcd	C 74.97	H 7.19	N 8.33
(336.4)	found	C 74.34	H 7.28	N 8.14

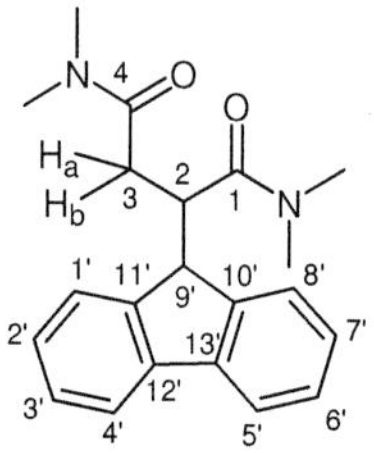

^{1}H NMR (500 MHz, CDCl$_3$): δ = 1.54 (dd, $^3J_{2,3Ha}$ = 2.8, $^2J_{3Ha,Hb}$ = 16.1 Hz, 1 H, 3-H$_a$), 2.69 (s, 3 H, NCH$_3$), 2.72 (dd, $^3J_{2,3}$ = 10.7, $^2J_{3a,3b}$ = 16.2 Hz, 1 H, 3-H$_b$), 2.77 (s, 3 H, NCH$_3$), 3.10 (s, 3 H, NCH$_3$), 3.29 (s, 3 H, NCH$_3$), 3.99 (ddd, $^3J_{3a,2}$ = 2.8, $^3J_{2,2'}$ = 4.0, $^3J_{2,3b}$ = 10.6 Hz, 1 H, 2-H), 4.08 (d, $^3J_{2,2'}$ = 4.1 Hz, 1 H, 9'-H), 7.28–7.38 (m, 4 H), 7.53 („d", J = 7.5 Hz, 1 H), 7.60 („d", J = 7.6 Hz, 1 H), 7.73 („d", J = 7.6 Hz, 2 H).

^{13}C NMR (125 MHz, CDCl$_3$): δ = 30.6 (t, C-3), 35.2 (q, NCH$_3$), 36.1 (q, NCH$_3$), 36.9 (q, NCH$_3$), 37.5 (q, NCH$_3$), 40.1 (d, C-2), 47.6 (d, C-9'), 119.7 (d), 119.8 (d), 124.1 (d), 126.4 (d), 127.0 (d), 127.1 (d), 127.4 (d), 127.6 (d), 141.2 and 141.2 (2 s, C-10', C-11'), 143.6 and 144.8 (2 s, C-12', C-13'), 171.6 (s, C=O), 173.9 (s, C=O).

The assignment of signals has been done by means of C,H COSY experiments. The structure of the succinic diamide **58** has been proven by means of X-ray crystal structure analysis.

Experiment 59 (AB 297)

A solution of the *N*-trimethylsilylimine **38** was prepared according to TP 7:

Fluorenone 360 mg, 2.0 mmol

LiHMDS ca. 2.2 mmol, 2.2 mL of ca. 1.0 M THF solution

A solution of the lithium enolate **56** was prepared as described in TP 12:

Diisopropylamine 0.34 mL, 242 mg, 2.4 mmol

n-Butyllithium .1.50 mL of ca. 1.58 M in hexane, ca. 2.4 mmol

N,N-Dimethylacetamide 209 mg, 0.22 mL, 2.4 mmol

The solution of the imine **38** was added to the solution of the lithium enolate **56** at −60 °C followed by stirring for 16 h at −30 °C. A colourless solid formed. The reaction was quenched by addition of water (5 mL), the aqueous phase was separated and the solid was extracted with ethyl acetate (4 ×, 50 mL). Drying and concentration of the organic layer gave 548 mg of a yellowish solid, which was purified on a silica gel (12 g, column 9 cm ×2.5 cm), eluting with MeOH/CH_2Cl_2 (3:97 to 5:95, v/v).

The main fraction, a solid, 404 mg, was recrystallized from ethyl acetate/petroleum ether yielding the spectroscopically and analytically pure succinic diamide **57** as a pale-yellow solid (290 mg, 60 % based on *N,N*-dimethylacetamide, 45 % based on fluorenone); m. p. 183–187 °C (179–183 °C, Exp. 58).

$C_{21}H_{24}N_2O_2$	calcd	C 74.97	H 7.19	N 8.33
(336.4)	found	C 74.97	H 7.19	N 8.20

NMR data of the diamide **57** were in accordance with those from the Exp. 58.

The minor fraction was a yellow solid, 80 mg ("15 %"). NMR data showed the presence of the crude β-amino ester **58** (stated on the basis of comparison with the NMR data of the following experiment 60).

Experiment 60 (AB 302)

A solution of the *N*-trimethylsilylimine **38** was prepared according to TP 7:

Fluorenone 360 mg, 2.0 mmol

LiHMDS 2.2 mL of ca. 1.0 M THF solution, ca. 2.2 mmol

The solution of the lithium enolate **56** was prepared as described in the TP 12:

Diisopropylamine 0.22 mL, 304 mg, 3.0 mmol

n-Butyllithium 1.9 mL of ca. 1.58 M in hexane, ca. 3.0 mmol

N,N-Dimethylacetamide 0.28 mL, 262 mg, 3.0 mmol

The solution of the imine **38** was added to the solution of the lithium enolate **56** at –50 °C followed by stirring the mixture for 7 h at this temperature. The reaction was quenched by addition of water (10 mL), the aqueous phase was separated and extracted with ethyl acetate (2 × 30 mL). The organic layer was washed with saturated aqueous NaCl solution, dried over MgSO$_4$ and concentrated *in vacuo* (0.1 mbar) giving 594 mg of a yellowish oil, which was purified on a SiO$_2$ column (18 g, 13 cm ×2.5 cm), eluting with MeOH/CH$_2$Cl$_2$ (0:100 to 3:97 to 5:95, v/v).

The minor fraction, an orange oil (124 mg, "23 %"), consisted of impure succinic diamide **57** in a mixture with β-amino ester **58**, according to ^{1}H and ^{13}C NMR data (**57** as the minor product).

The main fraction yielded the spectroscopically pure β-amino ester **58** as a colourless, viscous oil, 408 mg, 77 %.

IR (neat): ν =.3362 (w), 2928 (bw, NH$_2$), 1632 (s, C=O), 1447 (m), 1396 (s), 767 (s), 736 (vs) cm^{-1}.

C$_{17}$H$_{18}$N$_2$O	calcd	C 76.66	H 6.81	N 10.52
(266.3)	found	C 75.19	H 6.99	N 10.16

MS ESI (m/z) calc. for C$_{17}$H$_{18}$N$_2$O 266.14; found 267.15 [M+H]$^+$; 250.12 [M-NH$_3$]

HRMS: [M+H]$^+$ C$_{17}$H$_{18}$N$_2$O, calcd.: 267.1497, found: 267.1492

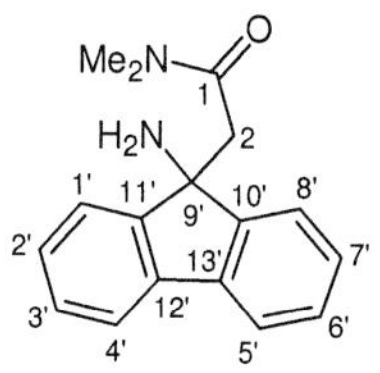

^{1}H NMR (500.1 MHz, CDCl$_3$): δ = 2.65 [2 s (not separated), 5 H together, NCH$_3$ and NH$_2$), 2.74 (s, 2 H, 2-H), 2.97 (s, 3 H, NCH$_3$), 7.27–7.37 (m, 4 H), 7.65–7.72 (m, 4 H).

^{13}C NMR (125.8 MHz, CDCl$_3$): δ = 35.4 (q, NCH$_3$), 37.6 (t, C-2), 42.6 (q, NCH$_3$), 64.3 (s, C-9'), 119.9 (d, C-4', C-5'), 124.3 (d, C-1', C-8'), 127.9 (d, C-3', C-6'), 128.3 (d, C-2', C-7'), 138.8 (s, C-10', C-11'), 151.3 (s, C-12', C-13'), 171.2 (s, C-1).

Experiment 61 (AB 323)

Attempted hydrolysis of the dimethylacetamide **58**; preparation of

2-(9-amino-9*H*-fluoren-9-yl)-*N*,*N*-dimethylacetamide
hydrochloride (59)

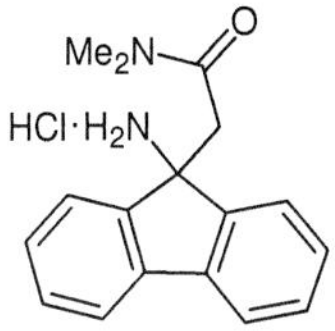

59

In close analogy to a known procedure,[131] 2-(9-amino-9*H*-fluoren-9-yl)-*N*,*N*-dimethylacetamide (**59**) (92 mg, 0.35 mmol) was suspended in a 80 % aqueous acetic acid (2.5 mL) and heated at 100 °C for 24 h. Then concentrated hydrochloric acid (0.1 mL) was added, and heating was continued for an additional 20 h. The volatiles were removed *in vacuo*, and the residual yellowish solid was washed with ether. The resulting colourless solid (90 mg, 90 %) was the title compound, spectroscopically pure and analytically almost pure, m. p. 237–240 °C (decomposition).

C$_{17}$H$_{19}$ClN$_2$O	calcd	C 66.29	H 6.34	N 9.03	Cl 11.69
(302.8)	found	C 67.43	H 6.32	N 9.25	Cl 11.71

IR (neat): ν =.2893 (b, m), 2747 (b, m), 2677 (b, m),2080 (b, w), 1602 (s, C=O), 1451 (s), 1400 (s), 764 (vs), 735 (vs) cm^{-1}.

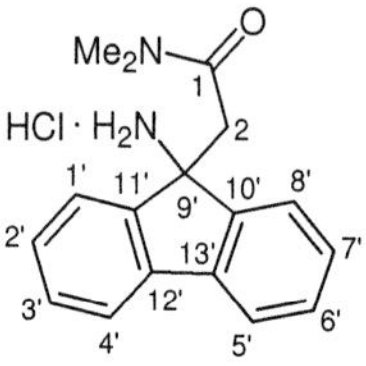

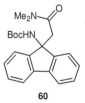

^{1}H NMR (250.1 MHz, CDCl$_3$): δ = 2.91 (s, 3 H, NCH$_3$), 3.07 (s, 3 H, NCH$_3$), 3.19 (s, 2 H, 2-H), 7.27–7.37 (m, 4 H), 7.65–7.72 (m, 4 H).

^{13}C NMR (62.9 MHz, CDCl$_3$): δ = 35.8 (q, NCH$_3$), 37.8 (q, NCH$_3$), 39.2 (t, C-2), 64.3 (s, C-9'), 122.1 (d, C-4', C-5'), 125.8 (d, C-1', C-8'), 129.8 (d, C-3', C-6'), 131.8 (d, C-2', C-7'), 141.4.8 (s, C-10', C-11'), 144.1 (s, C-12', C-13'), 171.6 (s, C-1).

Experiment 62 (AB 326)

2-(9-*tert*-Butoxycarbonylamino -9*H*-fluoren-9-yl)-*N,N*-dimethylacetamide (60)

60

In accordance with a literature procedure,[114] 2-(9-amino-9*H*-fluoren-9-yl)-*N,N*-dimethylacetamide (**59**, 243 mg, 0.99 mmol) was dissolved in acetonitrile (5 mL) and treated with di-*tert*-butyl dicarbonate (215 mg, 0.99 mmol). After stirring for 2 h the volatiles were removed *in vacuo*. The crude product, a yellowish semisolid, was purified by a flash-chromatography on silica gel (5 g, column 4.5 cm × 1.5 cm), eluting with ethyl acetate/petroleum ether (50:50, v/v). The *N*-Boc derivative **60** was obtained as a colourless, analytically pure foam (284 mg, 78 %), m. p. 124–125 °C.

IR (neat): ν = 3299 (w), 2675 (w), 1972 (w), 1700 (vs, C=O), 1623 (vs, C=O), 1420 (m), 1360 (s), 1165 (s), 1088 (s), 767 (s), 736 (vs) cm^{-1}.

C$_{22}$H$_{26}$N$_2$O$_3$	found	C 71.96	H 7.10	N 7.59
(366.5)	calcd	C 72.11	H 7.15	N 7.64

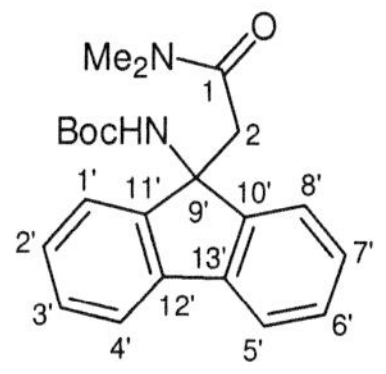

^{1}H NMR (250.1 MHz, CDCl$_3$): δ = 1.05 [bs, 9 H, C(CH$_3$)$_3$], 2.37 (s, 3 H, NCH$_3$), 2.61 (s, 2 H, 2-H), 2.97 (s, 3 H, NCH$_3$), 7.27–7.37 (m, 4 H), 7.65–7.72 (m, 4 H).

^{13}C NMR (62.9 MHz, CDCl$_3$): δ = 27.9 [q, C(*C*H$_3$)$_3$], 35.5 (q, NCH$_3$), 37.8 (q, NCH$_3$), 41.5 (t, C-2), 64.3 (s, C-9'), 78.9 [s, *C*(CH$_3$)$_3$], 120.2 (d, C-4', C-5'), 122.3 (d, C-1', C-8'), 127.5 (d, C-3', C-6'), 128.2 (d, C-2', C-7'), 139.0 (s, C-10', C-11'), 148.7 (s, C-12', C-13'), 153.9 (s, C, N*C*OOR), 170.7 (s, C-1).

8   APPENDIX

8.1   Crystal structures

8.1.1   (2S,3S)-3-(1-Adamantyl)-3-[N-(Z)-benzylidene-N-oxyamino]-1,2-isopropylidenepropane-1,2-diol (14)

$C_{23}H_{31}NO_3$
orthorhombic, $P2_12_12_1$
a =   6.3870(13) Å
b = 14.434(2) Å
c = 22.105(4) Å
α = 90 °
β = 90 °
γ = 90 °
V = 2037.9(6) Å^3

Z = 4, $R(F)$ = 0.0567
$R_w(F^2)$= 0.1053
Crystal size: 0.5 x 0.25 x 0.25 mm
Calculated density: 1.204 g/cm^3
2θ-Range for data collection: 1.68 - 25.00 °
Independent reflections: 3587
Observed reflections: 2320
Contributed reflections to refinement: 3587
Refined parameters: 245

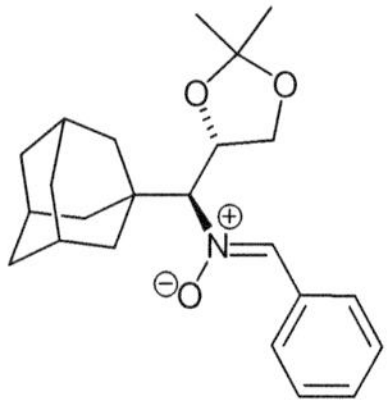

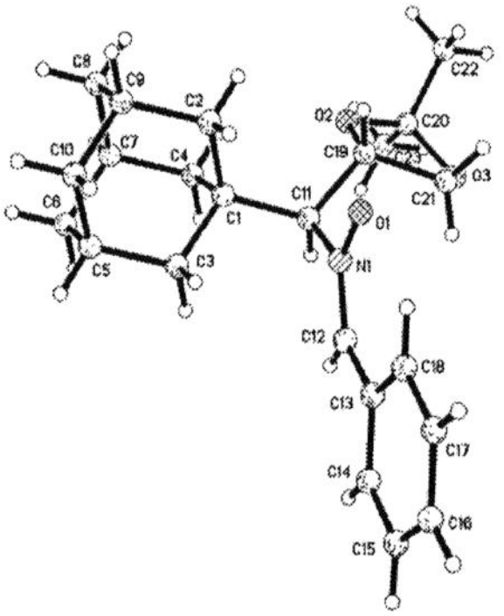

Elemental cell; View along the a-axis (a), b-axis (b) and c-axis (c):

(a)

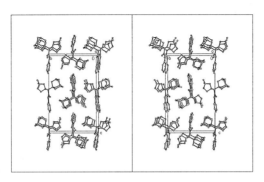

(b)

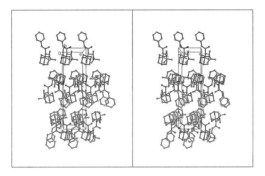

(c)

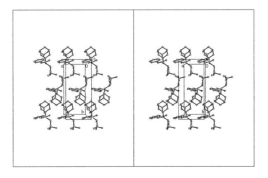

Bond lengths [Å] and angles [°]

C(1)-C(2)	1.520(4)	C(3)-C(1)-C(11)	109.4(2)
C(1)-C(3	1.528(4)	C(4)-C(1)-C(11)	109.0(2)
C(1)-C(4	1.537(4)	O(1)-N(1)-C(12)	124.2(2)
C(1)-C(11)	1.562(4)	O(1)-N(1)-C(11)	114.5(2)
O(1)-N(1)	1.292(3)	C(12)-N(1)-C(11)	121.3(3)
N(1)-C(12)	1.304(4)	C(20)-O(2)-C(19)	109.6(3)
N(1)-C(11)	1.493(3)	C(1)-C(2)-C(9)	110.3(3)
O(2)-C(20)	1.402(4)	C(1)-C(2)-H(2A)	109.6
O(2)-C(19)	1.420(4)	C(9)-C(2)-H(2A)	109.6
C(2)-C(9)	1.531(4)	C(1)-C(2)-H(2B)	109.6
C(2)-H(2A)	0.9700	C(9)-C(2)-H(2B)	109.6
C(2)-H(2B)	0.9700	H(2A)-C(2)-H(2B)	108.1
C(3)-C(5)	1.530(5)	C(1)-C(3)-C(5)	110.8(3)
C(3)-H(3A)	0.9700	C(1)-C(3)-H(3A)	109.5
C(3)-H(3B)	0.9700	C(5)-C(3)-H(3A)	109.5
O(3)-C(21)	1.370(4)	C(1)-C(3)-H(3B)	109.5
O(3)-C(20)	1.402(4)	C(5)-C(3)-H(3B)	109.5
C(4)-C(7)	1.528(4)	H(3A)-C(3)-H(3B)	108.1
C(4)-H(4A)	0.9700	C(21)-O(3)-C(20)	109.0(3)
C(4)-H(4B)	0.9700	C(7)-C(4)-C(1)	110.7(2)
C(5)-C(10)	1.512(5)	C(7)-C(4)-H(4A)	109.5
C(5)-C(6)	1.519(4)	C(1)-C(4)-H(4A)	109.5
C(5)-H(5)	0.9800	C(7)-C(4)-H(4B)	109.5
C(6)-C(7)	1.527(5)	C(1)-C(4)-H(4B)	109.5
C(6)-H(6A)	0.9700	H(4A)-C(4)-H(4B)	108.1
C(6)-H(6B)	0.9700	C(10)-C(5)-C(6)	109.2(3)
C(7)-C(8)	1.523(5)	C(10)-C(5)-C(3)	109.2(3)
C(7)-H(7)	0.9800	C(6)-C(5)-C(3)	110.1(3)
C(8)-C(9)	1.518(4)	C(10)-C(5)-H(5)	109.4
C(8)-H(8A)	0.9700	C(6)-C(5)-H(5)	109.4
C(8)-H(8B)	0.9700	C(3)-C(5)-H(5)	109.4
C(9)-C(10)	1.516(5)	C(5)-C(6)-C(7)	108.9(3)
C(9)-H(9)	0.9800	C(5)-C(6)-H(6A)	109.9
C(10)-H(10A)	0.9700	C(7)-C(6)-H(6A)	109.9
C(10)-H(10B)	0.9700	C(5)-C(6)-H(6B)	109.9
C(11)-C(19)	1.532(5)	C(7)-C(6)-H(6B)	109.9
C(11)-H(11)	0.9800	H(6A)-C(6)-H(6B)	108.3
C(12)-C(13)	1.441(4)	C(8)-C(7)-C(6)	109.9(3)
C(12)-H(12)	0.9300	C(8)-C(7)-C(4)	109.1(3)
C(13)-C(18)	1.390(4)	C(6)-C(7)-C(4)	109.5(3)
C(13)-C(14)	1.397(4)	C(8)-C(7)-H(7)	109.4
C(14)-C(15)	1.381(4)	C(6)-C(7)-H(7)	109.4
C(14)-H(14)	0.9300	C(4)-C(7)-H(7)	109.4
C(15)-C(16)	1.366(5)	C(9)-C(8)-C(7)	109.2(3)
C(15)-H(15)	0.9300	C(9)-C(8)-H(8A)	109.8
C(16)-C(17)	1.364(5)	C(7)-C(8)-H(8A)	109.8
C(16)-H(16)	0.9300	C(9)-C(8)-H(8B)	109.8
C(17)-C(18)	1.376(4)	C(7)-C(8)-H(8B)	109.8
C(17)-H(17)	0.9300	H(8A)-C(8)-H(8B)	108.3
C(18)-H(18)	0.9300	C(10)-C(9)-C(8)	109.5(3)
C(19)-C(21)	1.532(4)	C(10)-C(9)-C(2)	110.0(3)
C(19)-H(19)	0.9800	C(8)-C(9)-C(2)	109.3(3)
C(20)-C(23)	1.443(7)	C(10)-C(9)-H(9)	109.4
C(20)-C(22)	1.489(8)	C(8)-C(9)-H(9)	109.4
C(21)-H(21A)	0.9700	C(2)-C(9)-H(9)	109.4
C(21)-H(21B)	0.9700	C(5)-C(10)-C(9)	109.8(3)
C(22)-H(22A)	0.9600	C(5)-C(10)-H(10A)	109.7
C(22)-H(22B)	0.9600	C(9)-C(10)-H(10A)	109.7
C(22)-H(22C)	0.9600	C(5)-C(10)-H(10B)	109.7
C(23)-H(23A)	0.9600	C(9)-C(10)-H(10B)	109.7
C(23)-H(23B)	0.9600	H(10A)-C(10)-H(10B)	108.2
C(23)-H(23C)	0.9600	N(1)-C(11)-C(19)	105.7(3)
		N(1)-C(11)-C(1)	110.7(2)
C(2)-C(1)-C(3)	108.6(2)	C(19)-C(11)-C(1)	117.4(3)
C(2)-C(1)-C(4)	108.7(2)	N(1)-C(11)-H(11)	107.6
C(3)-C(1)-C(4)	107.7(3)	C(19)-C(11)-H(11)	107.6
C(2)-C(1)-C(11)	113.3(2)	C(1)-C(11)-H(11)	107.6

N(1)-C(12)-C(13)	127.3(3)	C(5)-C(6)-C(7)-C(4)	59.8(4)
N(1)-C(12)-H(12)	116.3	C(1)-C(4)-C(7)-C(8)	59.5(4)
C(13)-C(12)-H(12)	116.3	C(1)-C(4)-C(7)-C(6)	-60.8(4)
C(18)-C(13)-C(14)	118.0(3)	C(6)-C(7)-C(8)-C(9)	59.5(3)
C(18)-C(13)-C(12)	124.6(3)	C(4)-C(7)-C(8)-C(9)	-60.6(3)
C(14)-C(13)-C(12)	117.3(3)	C(7)-C(8)-C(9)-C(10)	-59.3(3)
C(15)-C(14)-C(13)	120.6(4)	C(7)-C(8)-C(9)-C(2)	61.2(4)
C(15)-C(14)-H(14)	119.7	C(1)-C(2)-C(9)-C(10)	59.4(3)
C(13)-C(14)-H(14)	119.7	C(1)-C(2)-C(9)-C(8)	-60.8(4)
C(16)-C(15)-C(14)	120.2(4)	C(6)-C(5)-C(10)-C(9)	-61.0(3)
C(16)-C(15)-H(15)	119.9	C(3)-C(5)-C(10)-C(9)	59.4(4)
C(14)-C(15)-H(15)	119.9	C(8)-C(9)-C(10)-C(5)	60.5(3)
C(17)-C(16)-C(15)	119.8(3)	C(2)-C(9)-C(10)-C(5)	-59.6(4)
C(17)-C(16)-H(16)	120.1	O(1)-N(1)-C(11)-C(19)	49.1(3)
C(15)-C(16)-H(16)	120.1	C(12)-N(1)-C(11)-C(19)	-132.6(3)
C(16)-C(17)-C(18)	121.1(4)	O(1)-N(1)-C(11)-C(1)	-79.0(3)
C(16)-C(17)-H(17)	119.5	C(12)-N(1)-C(11)-C(1)	99.4(3)
C(18)-C(17)-H(17)	119.5	C(2)-C(1)-C(11)-N(1)	73.7(3)
C(17)-C(18)-C(13)	120.2(4)	C(3)-C(1)-C(11)-N(1)	-47.7(3)
C(17)-C(18)-H(18)	119.9	C(4)-C(1)-C(11)-N(1)	-165.2(3)
C(13)-C(18)-H(18)	119.9	C(2)-C(1)-C(11)-C(19)	-47.7(3)
O(2)-C(19)-C(21)	103.5(3)	C(3)-C(1)-C(11)-C(19)	-169.1(3)
O(2)-C(19)-C(11)	110.2(3)	C(4)-C(1)-C(11)-C(19)	73.4(3)
C(21)-C(19)-C(11)	113.4(3)	O(1)-N(1)-C(12)-C(13)	1.0(5)
O(2)-C(19)-H(19)	109.8	C(11)-N(1)-C(12)-C(13)	-177.2(3)
C(21)-C(19)-H(19)	109.8	N(1)-C(12)-C(13)-C(18)	-2.9(5)
C(11)-C(19)-H(19)	109.8	N(1)-C(12)-C(13)-C(14)	176.4(3)
O(3)-C(20)-O(2)	105.9(3)	C(18)-C(13)-C(14)-C(15)	-1.0(5)
O(3)-C(20)-C(23)	112.4(5)	C(12)-C(13)-C(14)-C(15)	179.6(3)
O(2)-C(20)-C(23)	109.2(5)	C(13)-C(14)-C(15)-C(16)	-0.8(6)
O(3)-C(20)-C(22)	108.6(5)	C(14)-C(15)-C(16)-C(17)	1.9(6)
O(2)-C(20)-C(22)	110.1(5)	C(15)-C(16)-C(17)-C(18)	-1.2(6)
C(23)-C(20)-C(22)	110.4(5)	C(16)-C(17)-C(18)-C(13)	-0.6(6)
O(3)-C(21)-C(19)	105.0(3)	C(14)-C(13)-C(18)-C(17)	1.7(5)
O(3)-C(21)-H(21A)	110.7	C(12)-C(13)-C(18)-C(17)	-178.9(3)
C(19)-C(21)-H(21A)	110.7	C(20)-O(2)-C(19)-C(21)	-4.4(6)
O(3)-C(21)-H(21B)	110.7	C(20)-O(2)-C(19)-C(11)	-126.0(4)
C(19)-C(21)-H(21B)	110.7	N(1)-C(11)-C(19)-O(2)	170.7(2)
H(21A)-C(21)-H(21B)	108.8	C(1)-C(11)-C(19)-O(2)	-65.3(4)
C(20)-C(22)-H(22A)	109.5	N(1)-C(11)-C(19)-C(21)	55.2(4)
C(20)-C(22)-H(22B)	109.5	C(1)-C(11)-C(19)-C(21)	179.3(3)
H(22A)-C(22)-H(22B)	109.5	C(21)-O(3)-C(20)-O(2)	-27.4(6)
C(20)-C(22)-H(22C)	109.5	C(21)-O(3)-C(20)-C(23)	-146.6(5)
H(22A)-C(22)-H(22C)	109.5	C(21)-O(3)-C(20)-C(22)	90.8(5)
H(22B)-C(22)-H(22C)	109.5	C(19)-O(2)-C(20)-O(3)	18.9(6)
C(20)-C(23)-H(23A)	109.5	C(19)-O(2)-C(20)-C(23)	140.2(5)
C(20)-C(23)-H(23B)	109.5	C(19)-O(2)-C(20)-C(22)	-98.4(5)
H(23A)-C(23)-H(23B)	109.5	C(20)-O(3)-C(21)-C(19)	24.2(6)
C(20)-C(23)-H(23C)	109.5	O(2)-C(19)-C(21)-O(3)	-12.0(5)
H(23A)-C(23)-H(23C)	109.5	C(11)-C(19)-C(21)-O(3)	107.5(4)
H(23B)-C(23)-H(23C)	109.5		

Torsion angles [°]

C(3)-C(1)-C(2)-C(9)	-58.3(3)
C(4)-C(1)-C(2)-C(9)	58.6(3)
C(11)-C(1)-C(2)-C(9)	179.9(2)
C(2)-C(1)-C(3)-C(5)	59.0(3)
C(4)-C(1)-C(3)-C(5)	-58.6(3)
C(11)-C(1)-C(3)-C(5)	-176.9(2)
C(2)-C(1)-C(4)-C(7)	-58.3(4)
C(3)-C(1)-C(4)-C(7)	59.2(4)
C(11)-C(1)-C(4)-C(7)	177.8(3)
C(1)-C(3)-C(5)-C(10)	-59.8(3)
C(1)-C(3)-C(5)-C(6)	60.1(3)
C(10)-C(5)-C(6)-C(7)	60.4(4)
C(3)-C(5)-C(6)-C(7)	-59.5(4)
C(5)-C(6)-C(7)-C(8)	-60.1(4)

8.1.2 2,3-*O*-cyclohexylidene-D-glyceraldehyde *N*-benzylnitrone (16)

$C_{16}H_{21}NO_3$ $Z = 2$, $R(F) = 0.062$
monoclinic, $P2_1$ $R_w(F^2) = 0.1513$
$a = 10.882(3)$ Å   Crystal size: 0.8 x 0.3 x 0.25 mm
$b = 5.4217(16)$ Å   Calculated density: 1.251 g/cm^3
$c = 13.168(3)$Å   2θ-Range for data collection: 1.64 – 26.99 °
$\alpha = 90.0°$ Independent reflections: 3182
$\beta = 109.80(2)°$ Observed reflections: 2460
$\gamma = 90.0°$ Contributed reflections to refinement : 3182
$V = 731.0(4)$ Å^3 Refined parameters: 182

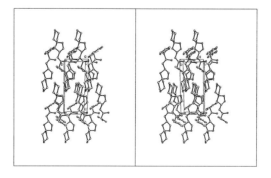

Elemental cell; View along the a-axis (a), b-axis (b) and c-axis (c):

a)

(b)

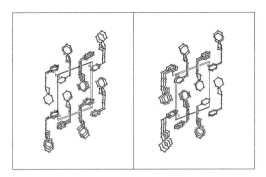

(c)

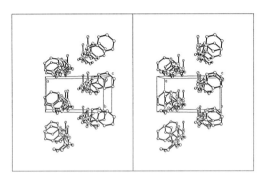

Bond lengths [Å] and angles [°]

N(1)-C(1)	1.289(4)
N(1)-O(1)	1.293(3)
N(1)-C(10)	1.483(3)
C(1)-C(2)	1.483(4)
C(1)-H(1)	0.9300
O(2)-C(2)	1.419(4)
O(2)-C(4)	1.440(3)
C(2)-C(3)	1.510(4)
C(2)-H(2)	0.9800
O(3)-C(4)	1.420(4)
O(3)-C(3)	1.427(4)
C(3)-H(3A)	0.9700
C(3)-H(3B)	0.9700
C(4)-C(9)	1.486(4)
C(4)-C(5)	1.511(4)
C(5)-C(6)	1.523(5)
C(5)-H(5A)	0.9700
C(5)-H(5B)	0.9700
C(6)-C(7)	1.481(6)
C(6)-H(6A)	0.9700
C(6)-H(6B)	0.9700
C(7)-C(8)	1.499(5)
C(7)-H(7A)	0.9700
C(7)-H(7B)	0.9700
C(8)-C(9)	1.523(5)
C(8)-H(8A)	0.9700
C(8)-H(8B)	0.9700
C(9)-H(9A)	0.9700
C(9)-H(9B)	0.9700
C(10)-C(11)	1.507(4)
C(10)-H(10A)	0.9700
C(10)-H(10B)	0.9700
C(11)-C(16)	1.380(4)
C(11)-C(12)	1.391(4)
C(12)-C(13)	1.376(4)
C(12)-H(12)	0.9300
C(13)-C(14)	1.367(5)
C(13)-H(13)	0.9300
C(14)-C(15)	1.372(5)
C(14)-H(14)	0.9300
C(15)-C(16)	1.382(4)
C(15)-H(15)	0.9300
C(16)-H(16)	0.9300
C(1)-N(1)-O(1)	123.3(2)

C(1)-N(1)-C(10)	120.7(2)	C(16)-C(11)-C(12)	118.5(3)
O(1)-N(1)-C(10)	116.0(2)	C(16)-C(11)-C(10)	119.7(3)
N(1)-C(1)-C(2)	120.8(3)	C(12)-C(11)-C(10)	121.9(3)
N(1)-C(1)-H(1)	119.6	C(13)-C(12)-C(11)	120.3(3)
C(2)-C(1)-H(1)	119.6	C(13)-C(12)-H(12)	119.9
C(2)-O(2)-C(4)	107.8(2)	C(11)-C(12)-H(12)	119.9
O(2)-C(2)-C(1)	108.4(2)	C(14)-C(13)-C(12)	120.6(3)
O(2)-C(2)-C(3)	102.3(2)	C(14)-C(13)-H(13)	119.7
C(1)-C(2)-C(3)	118.1(2)	C(12)-C(13)-H(13)	119.7
O(2)-C(2)-H(2)	109.2	C(13)-C(14)-C(15)	119.8(3)
C(1)-C(2)-H(2)	109.2	C(13)-C(14)-H(14)	120.1
C(3)-C(2)-H(2)	109.2	C(15)-C(14)-H(14)	120.1
C(4)-O(3)-C(3)	108.0(2)	C(14)-C(15)-C(16)	120.0(3)
O(3)-C(3)-C(2)	102.1(2)	C(14)-C(15)-H(15)	120.0
O(3)-C(3)-H(3A)	111.4	C(16)-C(15)-H(15)	120.0
C(2)-C(3)-H(3A)	111.4	C(11)-C(16)-C(15)	120.7(3)
O(3)-C(3)-H(3B)	111.4	C(11)-C(16)-H(16)	119.6
C(2)-C(3)-H(3B)	111.4	C(15)-C(16)-H(16)	119.6
H(3A)-C(3)-H(3B)	109.2		
O(3)-C(4)-O(2)	106.1(2)		
O(3)-C(4)-C(9)	111.2(3)	**Torsion angles [°]**	
O(2)-C(4)-C(9)	109.5(2)		
O(3)-C(4)-C(5)	108.1(3)	O(1)-N(1)-C(1)-C(2)	-3.9(4)
O(2)-C(4)-C(5)	110.4(2)	C(10)-N(1)-C(1)-C(2)	174.9(2)
C(9)-C(4)-C(5)	111.5(3)	C(4)-O(2)-C(2)-C(1)	-155.6(2)
C(4)-C(5)-C(6)	112.3(3)	C(4)-O(2)-C(2)-C(3)	-30.1(3)
C(4)-C(5)-H(5A)	109.1	N(1)-C(1)-C(2)-O(2)	175.1(2)
C(6)-C(5)-H(5A)	109.1	N(1)-C(1)-C(2)-C(3)	59.5(4)
C(4)-C(5)-H(5B)	109.1	C(4)-O(3)-C(3)-C(2)	-30.7(3)
C(6)-C(5)-H(5B)	109.1	O(2)-C(2)-C(3)-O(3)	36.8(3)
H(5A)-C(5)-H(5B)	107.9	C(1)-C(2)-C(3)-O(3)	155.7(3)
C(7)-C(6)-C(5)	111.8(4)	C(3)-O(3)-C(4)-O(2)	12.8(3)
C(7)-C(6)-H(6A)	109.2	C(3)-O(3)-C(4)-C(9)	-106.1(3)
C(5)-C(6)-H(6A)	109.2	C(3)-O(3)-C(4)-C(5)	131.3(3)
C(7)-C(6)-H(6B)	109.2	C(2)-O(2)-C(4)-O(3)	11.9(3)
C(5)-C(6)-H(6B)	109.2	C(2)-O(2)-C(4)-C(9)	131.9(3)
H(6A)-C(6)-H(6B)	107.9	C(2)-O(2)-C(4)-C(5)	-105.0(3)
C(6)-C(7)-C(8)	111.2(3)	O(3)-C(4)-C(5)-C(6)	70.8(5)
C(6)-C(7)-H(7A)	109.4	O(2)-C(4)-C(5)-C(6)	-173.6(4)
C(8)-C(7)-H(7A)	109.4	C(9)-C(4)-C(5)-C(6)	-51.7(5)
C(6)-C(7)-H(7B)	109.4	C(4)-C(5)-C(6)-C(7)	53.7(6)
C(8)-C(7)-H(7B)	109.4	C(5)-C(6)-C(7)-C(8)	-55.3(5)
H(7A)-C(7)-H(7B)	108.0	C(6)-C(7)-C(8)-C(9)	55.4(5)
C(7)-C(8)-C(9)	111.5(3)	O(3)-C(4)-C(9)-C(8)	-68.6(4)
C(7)-C(8)-H(8A)	109.3	O(2)-C(4)-C(9)-C(8)	174.6(3)
C(9)-C(8)-H(8A)	109.3	C(5)-C(4)-C(9)-C(8)	52.1(4)
C(7)-C(8)-H(8B)	109.3	C(7)-C(8)-C(9)-C(4)	-54.3(5)
C(9)-C(8)-H(8B)	109.3	C(1)-N(1)-C(10)-C(11)	112.3(3)
H(8A)-C(8)-H(8B)	108.0	O(1)-N(1)-C(10)-C(11)	-68.9(3)
C(4)-C(9)-C(8)	112.5(3)	N(1)-C(10)-C(11)-C(16)	-83.2(3)
C(4)-C(9)-H(9A)	109.1	N(1)-C(10)-C(11)-C(12)	97.1(3)
C(8)-C(9)-H(9A)	109.1	C(16)-C(11)-C(12)-C(13)	-1.3(4)
C(4)-C(9)-H(9B)	109.1		
C(8)-C(9)-H(9B)	109.1	C(10)-C(11)-C(12)-C(13)	178.4(3)
H(9A)-C(9)-H(9B)	107.8	C(11)-C(12)-C(13)-C(14)	0.8(5)
N(1)-C(10)-C(11)	111.8(2)	C(12)-C(13)-C(14)-C(15)	0.8(5)
N(1)-C(10)-H(10A)	109.3	C(13)-C(14)-C(15)-C(16)	-1.8(5)
C(11)-C(10)-H(10A)	109.3	C(12)-C(11)-C(16)-C(15)	0.3(4)
N(1)-C(10)-H(10B)	109.3		
C(11)-C(10)-H(10B)	109.3	C(10)-C(11)-C(16)-C(15)	-179.4(3)
H(10A)-C(10)-H(10B)	107.9	C(14)-C(15)-C(16)-C(11)	1.2(5)

8.1.3 (2*S*,3*S*)-3-(1-Adamantyl)-3-aminopropane-1,2-diol hydrochloride (21)

$C_{13}H_{26}ClNO_3$
orthorhombic, P2(1)2(1)2(1)
a = 6.532(2) Å
b = 6.7460(17) Å
c = 32.690(9) Å
α = 90 °
β = 90 °
γ = 90 °
V = 1440.4(7) Å^3
Z = 4, $R(F)$ = 0.0771

$R_w(F^2)$ = 0.1697
Crystal size: 0.4 x 0.25 x 0.15 mm
Calculated density: 1.290 g/cm^3
2θ-Range for data collection: 2.49–24.97 °
Independent reflections: 2520
Observed reflections: 1831
Contributed reflections to
refinement: 2520
Refined parameters: 174

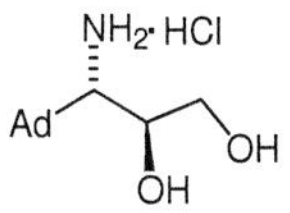

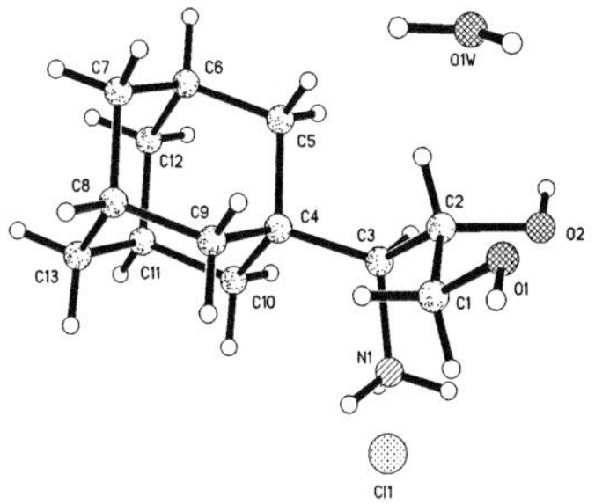

Elemental cell; View along the a-axis (a), b-axis (b) and c-axis (c):

(a)

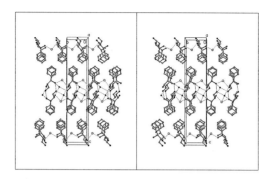

b)

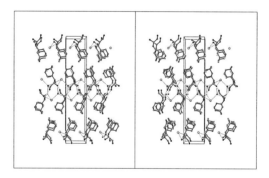

c)

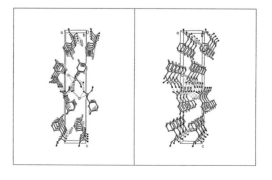

Bond lengths [Å] and angles [°]

O(1)-C(1)	1.420(6)	C(2)-C(3)-C(4)	119.0(4)
O(1)-H(1)	0.8200	N(1)-C(3)-H(3)	105.6
N(1)-C(3)	1.524(6)	C(2)-C(3)-H(3)	105.6
N(1)-H(1C)	0.8900	C(4)-C(3)-H(3)	105.6
N(1)-H(1D)	0.8900	C(9)-C(4)-C(3)	114.8(4)
N(1)-H(1E)	0.8900	C(9)-C(4)-C(10)	107.6(5)
C(1)-C(2)	1.511(8)	C(3)-C(4)-C(10)	111.0(4)
C(1)-H(1A)	0.9700	C(9)-C(4)-C(5)	108.0(5)
C(1)-H(1B)	0.9700	C(3)-C(4)-C(5)	107.4(4)
O(2)-C(2)	1.425(7)	C(10)-C(4)-C(5)	107.8(4)
O(2)-H(2A)	0.8200	C(6)-C(5)-C(4)	111.3(5)
C(2)-C(3)	1.536(6)	C(6)-C(5)-H(5A)	109.4
C(2)-H(2)	0.9800	C(4)-C(5)-H(5A)	109.4
C(3)-C(4)	1.536(6)	C(6)-C(5)-H(5B)	109.4
C(3)-H(3)	0.9800	C(4)-C(5)-H(5B)	109.4
C(4)-C(9)	1.532(7)	H(5A)-C(5)-H(5B)	108.0
C(4)-C(10)	1.545(7)	C(5)-C(6)-C(7)	109.7(6)
C(4)-C(5)	1.546(7)	C(5)-C(6)-C(12)	110.0(5)
C(5)-C(6)	1.519(9)	C(7)-C(6)-C(12)	108.9(6)
C(5)-H(5A)	0.9700	C(5)-C(6)-H(6)	109.4
C(5)-H(5B)	0.9700	C(7)-C(6)-H(6)	109.4
C(6)-C(7)	1.526(12)	C(12)-C(6)-H(6)	109.4
C(6)-C(12)	1.539(8)	C(6)-C(7)-C(8)	108.8(5)
C(6)-H(6)	0.9800	C(6)-C(7)-H(7A)	109.9
C(7)-C(8)	1.536(10)	C(8)-C(7)-H(7A)	109.9
C(7)-H(7A)	0.9700	C(6)-C(7)-H(7B)	109.9
C(7)-H(7B)	0.9700	C(8)-C(7)-H(7B)	109.9
C(8)-C(9)	1.520(8)	H(7A)-C(7)-H(7B)	108.3
C(8)-C(13)	1.522(10)	C(9)-C(8)-C(13)	111.1(6)
C(8)-H(8)	0.9800	C(9)-C(8)-C(7)	110.5(6)
C(9)-H(9A)	0.9700	C(13)-C(8)-C(7)	108.1(5)
C(9)-H(9B)	0.9700	C(9)-C(8)-H(8)	109.1
C(10)-C(11)	1.533(8)	C(13)-C(8)-H(8)	109.1
C(10)-H(10A)	0.9700	C(7)-C(8)-H(8)	109.1
C(10)-H(10B)	0.9700	C(8)-C(9)-C(4)	110.1(4)
C(11)-C(13)	1.496(11)	C(8)-C(9)-H(9A)	109.6
C(11)-C(12)	1.519(9)	C(4)-C(9)-H(9A)	109.6
C(11)-H(11)	0.9800	C(8)-C(9)-H(9B)	109.6
C(12)-H(12A)	0.9700	C(4)-C(9)-H(9B)	109.6
C(12)-H(12B)	0.9700	H(9A)-C(9)-H(9B)	108.2
C(13)-H(13A)	0.9700	C(11)-C(10)-C(4)	111.0(5)
C(13)-H(13B)	0.9700	C(11)-C(10)-H(10A)	109.4
O(1W)-H(1W)	0.896(10)	C(4)-C(10)-H(10A)	109.4
O(1W)-H(2W)	0.896(10)	C(11)-C(10)-H(10B)	109.4
		C(4)-C(10)-H(10B)	109.4
C(1)-O(1)-H(1)	109.5	H(10A)-C(10)-H(10B)	108.0
C(3)-N(1)-H(1C)	109.5	C(13)-C(11)-C(12)	110.7(6)
C(3)-N(1)-H(1D)	109.5	C(13)-C(11)-C(10)	109.1(6)
H(1C)-N(1)-H(1D)	109.5	C(12)-C(11)-C(10)	110.0(5)
C(3)-N(1)-H(1E)	109.5	C(13)-C(11)-H(11)	109.0
H(1C)-N(1)-H(1E)	109.5	C(12)-C(11)-H(11)	109.0
H(1D)-N(1)-H(1E)	109.5	C(10)-C(11)-H(11)	109.0
O(1)-C(1)-C(2)	112.0(4)	C(11)-C(12)-C(6)	108.3(4)
O(1)-C(1)-H(1A)	109.2	C(11)-C(12)-H(12A)	110.0
C(2)-C(1)-H(1A)	109.2	C(6)-C(12)-H(12A)	110.0
O(1)-C(1)-H(1B)	109.2	C(11)-C(12)-H(12B)	110.0
C(2)-C(1)-H(1B)	109.2	C(6)-C(12)-H(12B)	110.0
H(1A)-C(1)-H(1B)	107.9	H(12A)-C(12)-H(12B)	108.4
C(2)-O(2)-H(2A)	109.5	C(11)-C(13)-C(8)	109.3(5)
O(2)-C(2)-C(1)	106.5(4)	C(11)-C(13)-H(13A)	109.8
O(2)-C(2)-C(3)	109.3(4)	C(8)-C(13)-H(13A)	109.8
C(1)-C(2)-C(3)	115.8(4)	C(11)-C(13)-H(13B)	109.8
O(2)-C(2)-H(2)	108.3	C(8)-C(13)-H(13B)	109.8
C(1)-C(2)-H(2)	108.3	H(13A)-C(13)-H(13B)	108.3
C(3)-C(2)-H(2)	108.3	H(1W)-O(1W)-H(2W)	108.2(17)
N(1)-C(3)-C(2)	107.7(4)		
N(1)-C(3)-C(4)	112.4(4)		

Torsion angles [°]

O(1)-C(1)-C(2)-O(2)	56.2(6)
O(1)-C(1)-C(2)-C(3)	178.0(4)
O(2)-C(2)-C(3)-N(1)	76.3(5)
C(1)-C(2)-C(3)-N(1)	-43.9(6)
O(2)-C(2)-C(3)-C(4)	-154.4(4)
C(1)-C(2)-C(3)-C(4)	85.4(6)
N(1)-C(3)-C(4)-C(9)	78.7(6)
C(2)-C(3)-C(4)-C(9)	-48.5(7)
N(1)-C(3)-C(4)-C(10)	-43.6(5)
C(2)-C(3)-C(4)-C(10)	-170.8(4)
N(1)-C(3)-C(4)-C(5)	-161.2(4)
C(2)-C(3)-C(4)-C(5)	71.6(5)
C(9)-C(4)-C(5)-C(6)	-58.8(6)
C(3)-C(4)-C(5)-C(6)	176.9(5)
C(10)-C(4)-C(5)-C(6)	57.2(6)
C(4)-C(5)-C(6)-C(7)	59.8(7)
C(4)-C(5)-C(6)-C(12)	-60.0(7)
C(5)-C(6)-C(7)-C(8)	-58.8(7)
C(12)-C(6)-C(7)-C(8)	61.6(7)
C(6)-C(7)-C(8)-C(9)	59.8(7)
C(6)-C(7)-C(8)-C(13)	-61.9(7)
C(13)-C(8)-C(9)-C(4)	59.3(7)
C(7)-C(8)-C(9)-C(4)	-60.6(7)
C(3)-C(4)-C(9)-C(8)	178.2(5)
C(10)-C(4)-C(9)-C(8)	-57.6(7)
C(5)-C(4)-C(9)-C(8)	58.5(7)
C(9)-C(4)-C(10)-C(11)	59.1(6)
C(3)-C(4)-C(10)-C(11)	-174.5(5)
C(5)-C(4)-C(10)-C(11)	-57.2(6)
C(4)-C(10)-C(11)-C(13)	-61.0(7)
C(4)-C(10)-C(11)-C(12)	60.6(7)
C(13)-C(11)-C(12)-C(6)	60.1(7)
C(10)-C(11)-C(12)-C(6)	-60.6(6)
C(5)-C(6)-C(12)-C(11)	60.4(6)
C(7)-C(6)-C(12)-C(11)	-59.8(7)
C(12)-C(11)-C(13)-C(8)	-61.5(7)
C(10)-C(11)-C(13)-C(8)	59.7(7)
C(9)-C(8)-C(13)-C(11)	-59.9(7)
C(7)-C(8)-C(13)-C(11)	61.4(7)

H-Bond lengths [Å] and angles [°]

2.8432 (0.0055)	O1 - N1_$1
2.2040	H1 - N1_$1
3.0471 (0.0040)	O2 - Cl1_$2
2.2915	H2A - Cl1_$2
2.8432 (0.0055)	N1 - O1_$3
2.0029	H1C - O1_$3
3.1462 (0.0045)	N1 - Cl1
2.2959	H1D - Cl1
2.8960 (0.0060)	N1 - O1W_$4
2.0532	H1E - O1W_$4
3.0772 (0.0048)	O1W - Cl1_$5
2.2086 (0.0217)	H1W - Cl1_$5
2.8723 (0.0056)	O1W - O2_$6
2.0566 (0.0287)	H2W - O2_$6
134.96	O1 - H1 - N1_$1
153.44	O2 - H2A - Cl1_$2
156.94	N1 - H1C - O1_$3
159.79	N1 - H1D - Cl1
157.60	N1 - H1E - O1W_$4
163.13 (6.15)	O1W-H1W-Cl1_$5
150.84 (5.11)	O1W-H2W-O2_$6

$ stands for atoms of neigbouring molecules

8.1.4 (S)-1-Adamantylglycine hydrochloride [(S)-24]

C$_{12}$H$_{20}$ClNO$_2$·1/2CH$_3$OH
monoclinic, C2
a = 12.446(4) Å
b = 7.337(2) Å
c = 16.318(5) Å
α = 90°
β = 110.75(2)°
γ = 90°
V = 1393.5(8) Å^3

Z = 4, R(F) = 0.0922
$R_w(F^2)$ = 0.2316
Crystal size: 0.5 x 0.4 x 0.2 mm
Calculated density: 1.248 g/cm^3
2θ-Range for data collection: 2.67–26.00
Independent reflections: 2713
Observed reflections: 2109
Contributed reflections to refinement: 2713
Refined parameters: 158

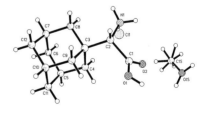

Elemental cell; View along the a-axis (a), b-axis (b) and c-axis (c):

a)

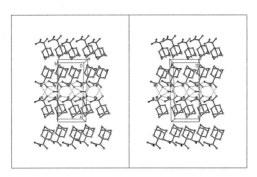

b)

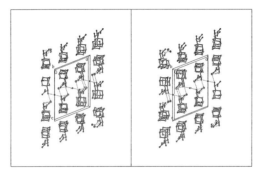

c)

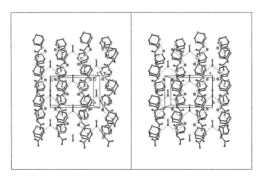

Bond lengths [Å] and angles [°]

N(1)-C(2)	1.479(9)	C(9)-C(3)-C(4)	107.7(6)
N(1)-H(1A)	0.8900	C(2)-C(3)-C(4)	112.6(6)
N(1)-H(1B)	0.8900	C(5)-C(4)-C(3)	108.9(7)
N(1)-H(1C)	0.8900	C(5)-C(4)-H(4A)	109.9
O(1)-C(1)	1.294(9)	C(3)-C(4)-H(4A)	109.9
O(1)-H(1)	0.8200	C(5)-C(4)-H(4B)	109.9
C(1)-O(2)	1.184(9)	C(3)-C(4)-H(4B)	109.9
C(1)-C(2)	1.544(10)	H(4A)-C(4)-H(4B)	108.3
C(2)-C(3)	1.542(9)	C(6)-C(5)-C(11)	112.4(9)
C(2)-H(2)	0.9800	C(6)-C(5)-C(4)	110.7(7)
C(3)-C(8)	1.528(10)	C(11)-C(5)-C(4)	108.3(8)
C(3)-C(9)	1.537(9)	C(6)-C(5)-H(5)	108.5
C(3)-C(4)	1.551(9)	C(11)-C(5)-H(5)	108.5
C(4)-C(5)	1.541(12)	C(4)-C(5)-H(5)	108.5
C(4)-H(4A)	0.9700	C(5)-C(6)-C(7)	108.2(8)
C(4)-H(4B)	0.9700	C(5)-C(6)-H(6A)	110.1
C(5)-C(6)	1.500(15)	C(7)-C(6)-H(6A)	110.1
C(5)-C(11)	1.526(14)	C(5)-C(6)-H(6B)	110.1
C(5)-H(5)	0.9800	C(7)-C(6)-H(6B)	110.1
C(6)-C(7)	1.547(15)	H(6A)-C(6)-H(6B)	108.4
C(6)-H(6A)	0.9700	C(12)-C(7)-C(8)	110.4(8)
C(6)-H(6B)	0.9700	C(12)-C(7)-C(6)	110.2(9)
C(7)-C(12)	1.515(15)	C(8)-C(7)-C(6)	108.8(8)
C(7)-C(8)	1.523(11)	C(12)-C(7)-H(7)	109.2
C(7)-H(7)	0.9800	C(8)-C(7)-H(7)	109.2
C(8)-H(8A)	0.9700	C(6)-C(7)-H(7)	109.2
C(8)-H(8B)	0.9700	C(7)-C(8)-C(3)	110.3(7)
C(9)-C(10)	1.529(12)	C(7)-C(8)-H(8A)	109.6
C(9)-H(9A)	0.9700	C(3)-C(8)-H(8A)	109.6
C(9)-H(9B)	0.9700	C(7)-C(8)-H(8B)	109.6
C(10)-C(12)	1.509(16)	C(3)-C(8)-H(8B)	109.6
C(10)-C(11)	1.552(15)	H(8A)-C(8)-H(8B)	108.1
C(10)-H(10)	0.9800	C(10)-C(9)-C(3)	110.4(7)
C(11)-H(11A)	0.9700	C(10)-C(9)-H(9A)	109.6
C(11)-H(11B)	0.9700	C(3)-C(9)-H(9A)	109.6
C(12)-H(12A)	0.9700	C(10)-C(9)-H(9B)	109.6
C(12)-H(12B)	0.9700	C(3)-C(9)-H(9B)	109.6
C(1S)-O(1S)	1.313(16)	H(9A)-C(9)-H(9B)	108.1
C(1S)-H(1S1)	0.9600	C(12)-C(10)-C(9)	110.3(8)
C(1S)-H(1S2)	0.9600	C(12)-C(10)-C(11)	110.1(10)
C(1S)-H(1S3)	0.9600	C(9)-C(10)-C(11)	108.0(8)
O(1S)-H(1S)	0.8200	C(12)-C(10)-H(10)	109.5
		C(9)-C(10)-H(10)	109.5
C(2)-N(1)-H(1A)	109.5	C(11)-C(10)-H(10)	109.5
C(2)-N(1)-H(1B)	109.5	C(5)-C(11)-C(10)	108.1(7)
H(1A)-N(1)-H(1B)	109.5	C(5)-C(11)-H(11A)	110.1
C(2)-N(1)-H(1C)	109.5	C(10)-C(11)-H(11A)	110.1
H(1A)-N(1)-H(1C)	109.5	C(5)-C(11)-H(11B)	110.1
H(1B)-N(1)-H(1C)	109.5	C(10)-C(11)-H(11B)	110.1
C(1)-O(1)-H(1)	109.5	H(11A)-C(11)-H(11B)	108.4
O(2)-C(1)-O(1)	125.0(7)	C(10)-C(12)-C(7)	109.6(8)
O(2)-C(1)-C(2)	121.8(6)	C(10)-C(12)-H(12A)	109.8
O(1)-C(1)-C(2)	113.2(6)	C(7)-C(12)-H(12A)	109.8
N(1)-C(2)-C(3)	113.6(6)	C(10)-C(12)-H(12B)	109.8
N(1)-C(2)-C(1)	104.3(5)	C(7)-C(12)-H(12B)	109.8
C(3)-C(2)-C(1)	114.8(6)	H(12A)-C(12)-H(12B)	108.2
N(1)-C(2)-H(2)	107.9	O(1S)-C(1S)-H(1S1)	109.5
C(3)-C(2)-H(2)	107.9	O(1S)-C(1S)-H(1S2)	109.5
C(1)-C(2)-H(2)	107.9	H(1S1)-C(1S)-H(1S2)	109.5
C(8)-C(3)-C(9)	108.8(6)	O(1S)-C(1S)-H(1S3)	109.5
C(8)-C(3)-C(2)	109.9(5)	H(1S1)-C(1S)-H(1S3)	109.5
C(9)-C(3)-C(2)	108.6(6)	H(1S2)-C(1S)-H(1S3)	109.5
C(8)-C(3)-C(4)	109.2(6)	C(1S)-O(1S)-H(1S)	109.5

Torsion angles [°]

O(2)-C(1)-C(2)-N(1)	-20.6(10)
O(1)-C(1)-C(2)-N(1)	159.3(7)
O(2)-C(1)-C(2)-C(3)	104.5(9)
O(1)-C(1)-C(2)-C(3)	-75.6(8)
N(1)-C(2)-C(3)-C(8)	-43.9(7)
C(1)-C(2)-C(3)-C(8)	-164.0(6)
N(1)-C(2)-C(3)-C(9)	-162.9(6)
C(1)-C(2)-C(3)-C(9)	77.1(7)
N(1)-C(2)-C(3)-C(4)	78.0(7)
C(1)-C(2)-C(3)-C(4)	-42.0(8)
C(8)-C(3)-C(4)-C(5)	-57.3(8)
C(9)-C(3)-C(4)-C(5)	60.8(9)
C(2)-C(3)-C(4)-C(5)	-179.6(7)
C(3)-C(4)-C(5)-C(6)	60.1(10)
C(3)-C(4)-C(5)-C(11)	-63.5(10)
C(11)-C(5)-C(6)-C(7)	59.2(9)
C(4)-C(5)-C(6)-C(7)	-62.0(10)
C(5)-C(6)-C(7)-C(12)	-59.1(10)
C(5)-C(6)-C(7)-C(8)	62.0(10)
C(12)-C(7)-C(8)-C(3)	59.7(10)
C(6)-C(7)-C(8)-C(3)	-61.3(10)
C(9)-C(3)-C(8)-C(7)	-58.0(9)
C(2)-C(3)-C(8)-C(7)	-176.8(7)
C(4)-C(3)-C(8)-C(7)	59.2(8)
C(8)-C(3)-C(9)-C(10)	57.7(9)
C(2)-C(3)-C(9)-C(10)	177.3(7)
C(4)-C(3)-C(9)-C(10)	-60.6(9)
C(3)-C(9)-C(10)-C(12)	-58.8(10)
C(3)-C(9)-C(10)-C(11)	61.6(10)
C(6)-C(5)-C(11)-C(10)	-58.8(11)
C(4)-C(5)-C(11)-C(10)	63.8(11)
C(12)-C(10)-C(11)-C(5)	58.0(10)
C(9)-C(10)-C(11)-C(5)	-62.5(11)
C(9)-C(10)-C(12)-C(7)	59.0(11)
C(11)-C(10)-C(12)-C(7)	-60.1(10)
C(8)-C(7)-C(12)-C(10)	-59.5(11)
C(6)-C(7)-C(12)-C(10)	60.6(10)

H-Bond lengths [Å] and angles [°]

2.9924 (0.0057)	O1 - Cl1_$1
2.1733	H1 - Cl1_$1
2.8155 (0.0077)	O1S - N1_$2
2.2355	H1S - N1_$2
3.1588 (0.0057)	N1 - Cl1_$2
2.3052	H1A - Cl1_$2
2.8155 (0.0077)	N1 - O1S_$3
1.9757	H1B - O1S_$3
3.1810 (0.0062)	N1 - Cl1
2.3243	H1C - Cl1
176.77	O1 - H1 - Cl1_$1
128.00	O1S - H1S - N1_$2
160.67	N1 - H1A - Cl1_$2
156.76	N1 - H1B - O1S_$3
161.52	N1 - H1C - Cl1

$ stands for atoms of neigbouring molecules

8.1.5   3-(9-Amino-9-fluorenyl)-azetidin-4-one-2-spiro-9'-fluorene (47)

C$_{28}$H$_{20}$N$_2$O
triclinic, P-1
a = 10.3075(16) Å
b = 14.308(2) Å
c = 14.841(3) Å
α = 109.971(13)°
β = 91.521(13)°
γ = 92.170(12)°
V = 2053.9(6) Å^3

Z = 4, $R(F)$ = 0.0567
$R_w(F^2)$ = 0.1256
Crystal size: 0.8 x 0.7 x 0.3 mm
Calculated density: 1.302 g/cm^3
2θ-Range for data collection: 1.46 – 27.50°
Independent reflections: 9433
Observed reflections: 6530
Contributed reflections to refinement: 9433
Refined parameters: 593

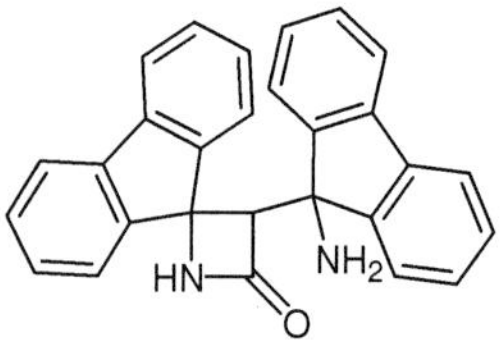

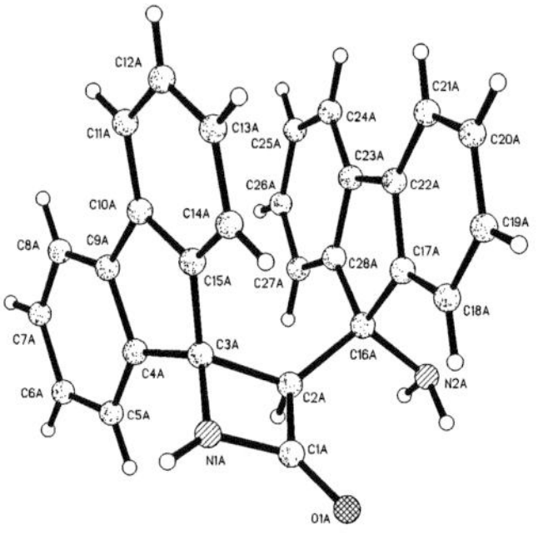

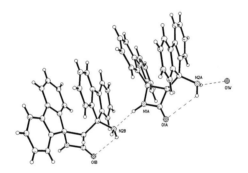

Structure

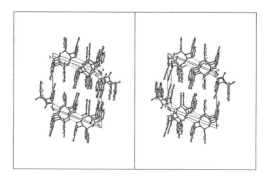

Elemental cell; View along the a-axis (a), b-axis (b) and c-axis (c):

a)

b)

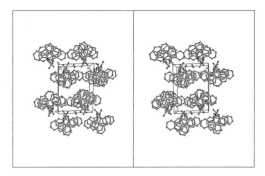

c)

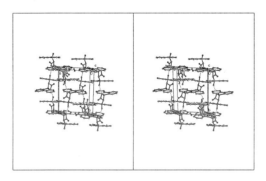

Bond lengths [Å] and angles [°]

N(1A)-C(1A)	1.348(2)	C(8A)-H(8A)	0.9300
N(1A)-C(3A)	1.476(2)	C(9A)-C(10A)	1.467(3)
N(1A)-H(1A)	0.95(2)	C(10A)-C(11A)	1.388(3)
O(1A)-C(1A)	1.215(2)	C(10A)-C(15A)	1.394(3)
C(1A)-C(2A)	1.535(2)	C(11A)-C(12A)	1.378(4)
C(2A)-C(16A)	1.534(2)	C(11A)-H(11A)	0.9300
C(2A)-C(3A)	1.592(2)	C(12A)-C(13A)	1.372(4)
C(2A)-H(2A)	0.9800	C(12A)-H(12A)	0.9300
C(3A)-C(4A)	1.513(3)	C(13A)-C(14A)	1.398(3)
C(3A)-C(15A)	1.517(3)	C(13A)-H(13A)	0.9300
C(4A)-C(5A)	1.385(3)	C(14A)-C(15A)	1.376(3)
C(4A)-C(9A)	1.399(3)	C(14A)-H(14A)	0.9300
C(5A)-C(6A)	1.389(3)	C(16A)-N(2A)	1.486(2)
C(5A)-H(5A)	0.9300	C(16A)-C(28A)	1.516(2)
C(6A)-C(7A)	1.377(4)	C(16A)-C(17A)	1.525(2)
C(6A)-H(6A)	0.9300	C(17A)-C(18A)	1.375(3)
C(7A)-C(8A)	1.381(4)	C(17A)-C(22A)	1.401(2)
C(7A)-H(7A)	0.9300	C(18A)-C(19A)	1.390(3)
C(8A)-C(9A)	1.397(3)	C(18A)-H(18A)	0.9300
		C(19A)-C(20A)	1.382(3)
		C(19A)-H(19A)	0.9300

C(20A)-C(21A)	1.372(3)	C(26B)-H(26B)	0.9300
C(20A)-H(20A)	0.9300	C(27B)-C(28B)	1.383(3)
C(21A)-C(22A)	1.385(3)	C(27B)-H(27B)	0.9300
C(21A)-H(21A)	0.9300	N(2B)-H(2B1)	0.95(2)
C(22A)-C(23A)	1.464(3)	N(2B)-H(2B2)	0.90(2)
C(23A)-C(24A)	1.388(3)	O(1W)-O(1W)#1	1.04(2)
C(23A)-C(28A)	1.397(3)		
C(24A)-C(25A)	1.385(3)	C(1A)-N(1A)-C(3A)	96.35(14)
C(24A)-H(24A)	0.9300	C(1A)-N(1A)-H(1A)	130.1(13)
C(25A)-C(26A)	1.368(4)	C(3A)-N(1A)-H(1A)	133.2(13)
C(25A)-H(25A)	0.9300	O(1A)-C(1A)-N(1A)	133.04(18)
C(26A)-C(27A)	1.386(3)	O(1A)-C(1A)-C(2A)	134.58(18)
C(26A)-H(26A)	0.9300	N(1A)-C(1A)-C(2A)	92.39(14)
C(27A)-C(28A)	1.382(3)	C(16A)-C(2A)-C(1A)	122.89(15)
C(27A)-H(27A)	0.9300	C(16A)-C(2A)-C(3A)	126.00(14)
N(2A)-H(2A1)	0.96(3)	C(1A)-C(2A)-C(3A)	84.66(13)
N(2A)-H(2A2)	0.86(3)	C(16A)-C(2A)-H(2A)	106.9
N(1B)-C(1B)	1.349(2)	C(1A)-C(2A)-H(2A)	106.9
N(1B)-C(3B)	1.472(2)	C(3A)-C(2A)-H(2A)	106.9
N(1B)-H(1B)	0.95(2)	N(1A)-C(3A)-C(4A)	121.70(16)
O(1B)-C(1B)	1.217(2)	N(1A)-C(3A)-C(15A)	116.23(16)
C(1B)-C(2B)	1.528(3)	C(4A)-C(3A)-C(15A)	102.79(16)
C(2B)-C(16B)	1.540(2)	N(1A)-C(3A)-C(2A)	85.47(13)
C(2B)-C(3B)	1.596(2)	C(4A)-C(3A)-C(2A)	112.66(15)
C(2B)-H(2B)	0.9800	C(15A)-C(3A)-C(2A)	118.48(15)
C(3B)-C(4B)	1.510(2)	C(5A)-C(4A)-C(9A)	121.0(2)
C(3B)-C(15B)	1.516(3)	C(5A)-C(4A)-C(3A)	129.71(19)
C(4B)-C(5B)	1.379(3)	C(9A)-C(4A)-C(3A)	109.04(18)
C(4B)-C(9B)	1.405(3)	C(4A)-C(5A)-C(6A)	118.9(2)
C(5B)-C(6B)	1.398(3)	C(4A)-C(5A)-H(5A)	120.5
C(5B)-H(5B)	0.9300	C(6A)-C(5A)-H(5A)	120.5
C(6B)-C(7B)	1.375(3)	C(7A)-C(6A)-C(5A)	120.1(3)
C(6B)-H(6B)	0.9300	C(7A)-C(6A)-H(6A)	119.9
C(7B)-C(8B)	1.388(3)	C(5A)-C(6A)-H(6A)	119.9
C(7B)-H(7B)	0.9300	C(6A)-C(7A)-C(8A)	121.6(2)
C(8B)-C(9B)	1.391(3)	C(6A)-C(7A)-H(7A)	119.2
C(8B)-H(8B)	0.9300	C(8A)-C(7A)-H(7A)	119.2
C(9B)-C(10B)	1.469(3)	C(7A)-C(8A)-C(9A)	118.9(2)
C(10B)-C(11B)	1.389(3)	C(7A)-C(8A)-H(8A)	120.5
C(10B)-C(15B)	1.398(3)	C(9A)-C(8A)-H(8A)	120.5
C(11B)-C(12B)	1.389(3)	C(8A)-C(9A)-C(4A)	119.3(2)
C(11B)-H(11B)	0.9300	C(8A)-C(9A)-C(10A)	131.8(2)
C(12B)-C(13B)	1.385(3)	C(4A)-C(9A)-C(10A)	108.89(19)
C(12B)-H(12B)	0.9300	C(11A)-C(10A)-C(15A)	120.0(2)
C(13B)-C(14B)	1.387(3)	C(11A)-C(10A)-C(9A)	131.1(2)
C(13B)-H(13B)	0.9300	C(15A)-C(10A)-C(9A)	108.75(18)
C(14B)-C(15B)	1.378(3)	C(12A)-C(11A)-C(10A)	118.8(3)
C(14B)-H(14B)	0.9300	C(12A)-C(11A)-H(11A)	120.6
C(16B)-N(2B)	1.486(2)	C(10A)-C(11A)-H(11A)	120.6
C(16B)-C(17B)	1.524(2)	C(13A)-C(12A)-C(11A)	121.1(2)
C(16B)-C(28B)	1.525(2)	C(13A)-C(12A)-H(12A)	119.5
C(17B)-C(18B)	1.380(3)	C(11A)-C(12A)-H(12A)	119.5
C(17B)-C(22B)	1.398(2)	C(12A)-C(13A)-C(14A)	120.7(3)
C(18B)-C(19B)	1.393(3)	C(12A)-C(13A)-H(13A)	119.6
C(18B)-H(18B)	0.9300	C(14A)-C(13A)-H(13A)	119.6
C(19B)-C(20B)	1.380(3)	C(15A)-C(14A)-C(13A)	118.2(2)
C(19B)-H(19B)	0.9300	C(15A)-C(14A)-H(14A)	120.9
C(20B)-C(21B)	1.380(3)	C(13A)-C(14A)-H(14A)	120.9
C(20B)-H(20B)	0.9300	C(14A)-C(15A)-C(10A)	121.0(2)
C(21B)-C(22B)	1.385(3)	C(14A)-C(15A)-C(3A)	129.7(2)
C(21B)-H(21B)	0.9300	C(10A)-C(15A)-C(3A)	109.35(18)
C(22B)-C(23B)	1.465(3)	N(2A)-C(16A)-C(28A)	109.86(15)
C(23B)-C(24B)	1.385(3)	N(2A)-C(16A)-C(17A)	108.70(15)
C(23B)-C(28B)	1.402(2)	C(28A)-C(16A)-C(17A)	101.94(14)
C(24B)-C(25B)	1.385(3)	N(2A)-C(16A)-C(2A)	109.61(14)
C(24B)-H(24B)	0.9300	C(28A)-C(16A)-C(2A)	110.44(14)
C(25B)-C(26B)	1.375(3)	C(17A)-C(16A)-C(2A)	115.99(15)
C(25B)-H(25B)	0.9300	C(18A)-C(17A)-C(22A)	120.54(18)
C(26B)-C(27B)	1.390(3)	C(18A)-C(17A)-C(16A)	129.53(18)

C(22A)-C(17A)-C(16A)	109.90(16)		C(9B)-C(8B)-H(8B)	120.8
C(17A)-C(18A)-C(19A)	118.7(2)		C(8B)-C(9B)-C(4B)	120.31(18)
C(17A)-C(18A)-H(18A)	120.6		C(8B)-C(9B)-C(10B)	130.81(19)
C(19A)-C(18A)-H(18A)	120.6		C(4B)-C(9B)-C(10B)	108.80(16)
C(20A)-C(19A)-C(18A)	120.4(2)		C(11B)-C(10B)-C(15B)	120.13(19)
C(20A)-C(19A)-H(19A)	119.8		C(11B)-C(10B)-C(9B)	131.64(19)
C(18A)-C(19A)-H(19A)	119.8		C(15B)-C(10B)-C(9B)	108.21(17)
C(21A)-C(20A)-C(19A)	121.2(2)		C(10B)-C(11B)-C(12B)	118.4(2)
C(21A)-C(20A)-H(20A)	119.4		C(10B)-C(11B)-H(11B)	120.8
C(19A)-C(20A)-H(20A)	119.4		C(12B)-C(11B)-H(11B)	120.8
C(20A)-C(21A)-C(22A)	118.8(2)		C(13B)-C(12B)-C(11B)	121.1(2)
C(20A)-C(21A)-H(21A)	120.6		C(13B)-C(12B)-H(12B)	119.4
C(22A)-C(21A)-H(21A)	120.6		C(11B)-C(12B)-H(12B)	119.4
C(21A)-C(22A)-C(17A)	120.22(18)		C(12B)-C(13B)-C(14B)	120.4(2)
C(21A)-C(22A)-C(23A)	130.78(18)		C(12B)-C(13B)-H(13B)	119.8
C(17A)-C(22A)-C(23A)	108.99(16)		C(14B)-C(13B)-H(13B)	119.8
C(24A)-C(23A)-C(28A)	120.61(19)		C(15B)-C(14B)-C(13B)	118.7(2)
C(24A)-C(23A)-C(22A)	130.94(19)		C(15B)-C(14B)-H(14B)	120.6
C(28A)-C(23A)-C(22A)	108.44(16)		C(13B)-C(14B)-H(14B)	120.6
C(25A)-C(24A)-C(23A)	117.7(2)		C(14B)-C(15B)-C(10B)	121.01(18)
C(25A)-C(24A)-H(24A)	121.1		C(14B)-C(15B)-C(3B)	129.10(18)
C(23A)-C(24A)-H(24A)	121.1		C(10B)-C(15B)-C(3B)	109.75(16)
C(26A)-C(25A)-C(24A)	121.7(2)		N(2B)-C(16B)-C(17B)	110.42(15)
C(26A)-C(25A)-H(25A)	119.1		N(2B)-C(16B)-C(28B)	108.48(14)
C(24A)-C(25A)-H(25A)	119.1		C(17B)-C(16B)-C(28B)	101.80(14)
C(25A)-C(26A)-C(27A)	121.0(2)		N(2B)-C(16B)-C(2B)	109.56(14)
C(25A)-C(26A)-H(26A)	119.5		C(17B)-C(16B)-C(2B)	110.49(14)
C(27A)-C(26A)-H(26A)	119.5		C(28B)-C(16B)-C(2B)	115.82(15)
C(28A)-C(27A)-C(26A)	118.2(2)		C(18B)-C(17B)-C(22B)	121.20(17)
C(28A)-C(27A)-H(27A)	120.9		C(18B)-C(17B)-C(16B)	128.45(17)
C(26A)-C(27A)-H(27A)	120.9		C(22B)-C(17B)-C(16B)	110.32(16)
C(27A)-C(28A)-C(23A)	120.76(18)		C(17B)-C(18B)-C(19B)	118.06(19)
C(27A)-C(28A)-C(16A)	128.53(18)		C(17B)-C(18B)-H(18B)	121.0
C(23A)-C(28A)-C(16A)	110.71(16)		C(19B)-C(18B)-H(18B)	121.0
C(16A)-N(2A)-H(2A1)	109.7(17)		C(20B)-C(19B)-C(18B)	120.8(2)
C(16A)-N(2A)-H(2A2)	104.8(18)		C(20B)-C(19B)-H(19B)	119.6
H(2A1)-N(2A)-H(2A2)	111(2)		C(18B)-C(19B)-H(19B)	119.6
C(1B)-N(1B)-C(3B)	96.10(15)		C(21B)-C(20B)-C(19B)	121.24(19)
C(1B)-N(1B)-H(1B)	130.4(15)		C(21B)-C(20B)-H(20B)	119.4
C(3B)-N(1B)-H(1B)	133.3(15)		C(19B)-C(20B)-H(20B)	119.4
O(1B)-C(1B)-N(1B)	132.2(2)		C(20B)-C(21B)-C(22B)	118.66(19)
O(1B)-C(1B)-C(2B)	134.64(19)		C(20B)-C(21B)-H(21B)	120.7
N(1B)-C(1B)-C(2B)	93.19(15)		C(22B)-C(21B)-H(21B)	120.7
C(1B)-C(2B)-C(16B)	121.85(15)		C(21B)-C(22B)-C(17B)	120.07(18)
C(1B)-C(2B)-C(3B)	84.41(13)		C(21B)-C(22B)-C(23B)	130.97(18)
C(16B)-C(2B)-C(3B)	125.84(14)		C(17B)-C(22B)-C(23B)	108.93(15)
C(1B)-C(2B)-H(2B)	107.4		C(24B)-C(23B)-C(28B)	120.57(18)
C(16B)-C(2B)-H(2B)	107.4		C(24B)-C(23B)-C(22B)	130.82(18)
C(3B)-C(2B)-H(2B)	107.4		C(28B)-C(23B)-C(22B)	108.60(16)
N(1B)-C(3B)-C(4B)	117.78(16)		C(25B)-C(24B)-C(23B)	118.4(2)
N(1B)-C(3B)-C(15B)	119.47(15)		C(25B)-C(24B)-H(24B)	120.8
C(4B)-C(3B)-C(15B)	102.44(14)		C(23B)-C(24B)-H(24B)	120.8
N(1B)-C(3B)-C(2B)	85.98(13)		C(26B)-C(25B)-C(24B)	121.3(2)
C(4B)-C(3B)-C(2B)	119.13(14)		C(26B)-C(25B)-H(25B)	119.4
C(15B)-C(3B)-C(2B)	112.60(15)		C(24B)-C(25B)-H(25B)	119.4
C(5B)-C(4B)-C(9B)	120.57(17)		C(25B)-C(26B)-C(27B)	120.7(2)
C(5B)-C(4B)-C(3B)	129.98(17)		C(25B)-C(26B)-H(26B)	119.7
C(9B)-C(4B)-C(3B)	109.43(16)		C(27B)-C(26B)-H(26B)	119.7
C(4B)-C(5B)-C(6B)	118.5(2)		C(28B)-C(27B)-C(26B)	118.8(2)
C(4B)-C(5B)-H(5B)	120.7		C(28B)-C(27B)-H(27B)	120.6
C(6B)-C(5B)-H(5B)	120.7		C(26B)-C(27B)-H(27B)	120.6
C(7B)-C(6B)-C(5B)	120.9(2)		C(27B)-C(28B)-C(23B)	120.28(17)
C(7B)-C(6B)-H(6B)	119.6		C(27B)-C(28B)-C(16B)	129.40(17)
C(5B)-C(6B)-H(6B)	119.6		C(23B)-C(28B)-C(16B)	110.31(15)
C(6B)-C(7B)-C(8B)	121.12(19)		C(16B)-N(2B)-H(2B1)	108.1(13)
C(6B)-C(7B)-H(7B)	119.4		C(16B)-N(2B)-H(2B2)	105.2(15)
C(8B)-C(7B)-H(7B)	119.4		H(2B1)-N(2B)-H(2B2)	109.9(19)
C(7B)-C(8B)-C(9B)	118.5(2)			
C(7B)-C(8B)-H(8B)	120.8			

Torsion angles [°]

C(3A)-N(1A)-C(1A)-O(1A)	171.6(2)		C(22A)-C(17A)-C(18A)-C(19A)	1.9(3)
C(3A)-N(1A)-C(1A)-C(2A)	-8.62(15)		C(16A)-C(17A)-C(18A)-C(19A)	-179.9(2)
O(1A)-C(1A)-C(2A)-C(16A)	-42.2(3)		C(17A)-C(18A)-C(19A)-C(20A)	-0.8(4)
N(1A)-C(1A)-C(2A)-C(16A)	137.97(17)		C(18A)-C(19A)-C(20A)-C(21A)	-0.4(4)
O(1A)-C(1A)-C(2A)-C(3A)	-172.2(2)		C(19A)-C(20A)-C(21A)-C(22A)	0.4(4)
N(1A)-C(1A)-C(2A)-C(3A)	7.98(14)		C(20A)-C(21A)-C(22A)-C(17A)	0.8(3)
C(1A)-N(1A)-C(3A)-C(4A)	122.24(18)		C(20A)-C(21A)-C(22A)-C(23A)	179.6(2)
C(1A)-N(1A)-C(3A)-C(15A)	-111.29(17)		C(18A)-C(17A)-C(22A)-C(21A)	-2.0(3)
C(1A)-N(1A)-C(3A)-C(2A)	8.33(15)		C(16A)-C(17A)-C(22A)-C(21A)	179.54(17)
C(16A)-C(2A)-C(3A)-N(1A)	-134.62(18)		C(18A)-C(17A)-C(22A)-C(23A)	178.99(17)
C(1A)-C(2A)-C(3A)-N(1A)	-7.29(13)		C(16A)-C(17A)-C(22A)-C(23A)	0.5(2)
C(16A)-C(2A)-C(3A)-C(4A)	102.8(2)		C(21A)-C(22A)-C(23A)-C(24A)	1.4(4)
C(1A)-C(2A)-C(3A)-C(4A)	-129.85(16)		C(17A)-C(22A)-C(23A)-C(24A)	-179.7(2)
C(16A)-C(2A)-C(3A)-C(15A)	-17.1(3)		C(21A)-C(22A)-C(23A)-C(28A)	180.0(2)
C(1A)-C(2A)-C(3A)-C(15A)	110.19(17)		C(17A)-C(22A)-C(23A)-C(28A)	-1.2(2)
N(1A)-C(3A)-C(4A)-C(5A)	-42.4(3)		C(28A)-C(23A)-C(24A)-C(25A)	-0.3(3)
C(15A)-C(3A)-C(4A)-C(5A)	-174.7(2)		C(22A)-C(23A)-C(24A)-C(25A)	178.1(2)
C(2A)-C(3A)-C(4A)-C(5A)	56.6(3)		C(23A)-C(24A)-C(25A)-C(26A)	0.1(3)
N(1A)-C(3A)-C(4A)-C(9A)	142.93(17)		C(24A)-C(25A)-C(26A)-C(27A)	0.4(4)
C(15A)-C(3A)-C(4A)-C(9A)	10.6(2)		C(25A)-C(26A)-C(27A)-C(28A)	-0.6(4)
C(2A)-C(3A)-C(4A)-C(9A)	-118.02(17)		C(26A)-C(27A)-C(28A)-C(23A)	0.4(3)
C(9A)-C(4A)-C(5A)-C(6A)	0.5(3)		C(26A)-C(27A)-C(28A)-C(16A)	-179.7(2)
C(3A)-C(4A)-C(5A)-C(6A)	-173.7(2)		C(24A)-C(23A)-C(28A)-C(27A)	0.0(3)
C(4A)-C(5A)-C(6A)-C(7A)	1.4(4)		C(22A)-C(23A)-C(28A)-C(27A)	-178.69(17)
C(5A)-C(6A)-C(7A)-C(8A)	-1.5(4)		C(24A)-C(23A)-C(28A)-C(16A)	-179.92(17)
C(6A)-C(7A)-C(8A)-C(9A)	-0.2(4)		C(22A)-C(23A)-C(28A)-C(16A)	1.4(2)
C(7A)-C(8A)-C(9A)-C(4A)	2.0(3)		N(2A)-C(16A)-C(28A)-C(27A)	63.9(2)
C(7A)-C(8A)-C(9A)-C(10A)	-177.3(2)		C(17A)-C(16A)-C(28A)-C(27A)	179.07(19)
C(5A)-C(4A)-C(9A)-C(8A)	-2.1(3)		C(2A)-C(16A)-C(28A)-C(27A)	-57.1(2)
C(3A)-C(4A)-C(9A)-C(8A)	173.09(18)		N(2A)-C(16A)-C(28A)-C(23A)	-116.14(17)
C(5A)-C(4A)-C(9A)-C(10A)	177.30(18)		C(17A)-C(16A)-C(28A)-C(23A)	-1.00(18)
C(3A)-C(4A)-C(9A)-C(10A)	-7.5(2)		C(2A)-C(16A)-C(28A)-C(23A)	122.83(16)
C(8A)-C(9A)-C(10A)-C(11A)	3.9(4)		C(3B)-N(1B)-C(1B)-O(1B)	-175.4(2)
C(4A)-C(9A)-C(10A)-C(11A)	-175.5(2)		C(3B)-N(1B)-C(1B)-C(2B)	4.61(15)
C(8A)-C(9A)-C(10A)-C(15A)	-179.8(2)		O(1B)-C(1B)-C(2B)-C(16B)	46.6(3)
C(4A)-C(9A)-C(10A)-C(15A)	0.8(2)		N(1B)-C(1B)-C(2B)-C(16B)	-133.43(17)
C(15A)-C(10A)-C(11A)-C(12A)	-1.3(3)		O(1B)-C(1B)-C(2B)-C(3B)	175.8(3)
C(9A)-C(10A)-C(11A)-C(12A)	174.6(2)		N(1B)-C(1B)-C(2B)-C(3B)	-4.25(14)
C(10A)-C(11A)-C(12A)-C(13A)	-2.2(4)		C(1B)-N(1B)-C(3B)-C(4B)	116.56(17)
C(11A)-C(12A)-C(13A)-C(14A)	3.0(4)		C(1B)-N(1B)-C(3B)-C(15B)	-118.17(18)
C(12A)-C(13A)-C(14A)-C(15A)	-0.2(3)		C(1B)-N(1B)-C(3B)-C(2B)	-4.42(15)
C(13A)-C(14A)-C(15A)-C(10A)	-3.3(3)		C(1B)-C(2B)-C(3B)-N(1B)	3.89(13)
C(13A)-C(14A)-C(15A)-C(3A)	178.11(19)		C(16B)-C(2B)-C(3B)-N(1B)	129.57(18)
C(11A)-C(10A)-C(15A)-C(14A)	4.1(3)		C(1B)-C(2B)-C(3B)-C(4B)	-115.83(17)
C(9A)-C(10A)-C(15A)-C(14A)	-172.64(18)		C(16B)-C(2B)-C(3B)-C(4B)	9.9(3)
C(11A)-C(10A)-C(15A)-C(3A)	-177.05(19)		C(1B)-C(2B)-C(3B)-C(15B)	124.22(16)
C(9A)-C(10A)-C(15A)-C(3A)	6.2(2)		C(16B)-C(2B)-C(3B)-C(15B)	-110.10(19)
N(1A)-C(3A)-C(15A)-C(14A)	33.1(3)		N(1B)-C(3B)-C(4B)-C(5B)	-34.8(3)
C(4A)-C(3A)-C(15A)-C(14A)	168.5(2)		C(15B)-C(3B)-C(4B)-C(5B)	-168.13(18)
C(2A)-C(3A)-C(15A)-C(14A)	-66.6(3)		C(2B)-C(3B)-C(4B)-C(5B)	66.9(3)
N(1A)-C(3A)-C(15A)-C(10A)	-145.60(17)		N(1B)-C(3B)-C(4B)-C(9B)	143.33(16)
C(4A)-C(3A)-C(15A)-C(10A)	-10.2(2)		C(15B)-C(3B)-C(4B)-C(9B)	10.04(19)
C(2A)-C(3A)-C(15A)-C(10A)	114.78(18)		C(2B)-C(3B)-C(4B)-C(9B)	-114.96(18)
C(1A)-C(2A)-C(16A)-N(2A)	73.4(2)		C(9B)-C(4B)-C(5B)-C(6B)	3.1(3)
C(3A)-C(2A)-C(16A)-N(2A)	-177.13(17)		C(3B)-C(4B)-C(5B)-C(6B)	-178.90(18)
C(1A)-C(2A)-C(16A)-C(28A)	-165.41(16)		C(4B)-C(5B)-C(6B)-C(7B)	-0.6(3)
C(3A)-C(2A)-C(16A)-C(28A)	-56.0(2)		C(5B)-C(6B)-C(7B)-C(8B)	-1.5(3)
C(1A)-C(2A)-C(16A)-C(17A)	-50.1(2)		C(6B)-C(7B)-C(8B)-C(9B)	1.0(3)
C(3A)-C(2A)-C(16A)-C(17A)	59.3(2)		C(7B)-C(8B)-C(9B)-C(4B)	1.6(3)
N(2A)-C(16A)-C(17A)-C(18A)	-62.0(2)		C(7B)-C(8B)-C(9B)-C(10B)	-174.7(2)
C(28A)-C(16A)-C(17A)-C(18A)	178.02(19)		C(5B)-C(4B)-C(9B)-C(8B)	-3.6(3)
C(2A)-C(16A)-C(17A)-C(18A)	62.0(3)		C(3B)-C(4B)-C(9B)-C(8B)	177.99(17)
N(2A)-C(16A)-C(17A)-C(22A)	116.23(17)		C(5B)-C(4B)-C(9B)-C(10B)	173.38(17)
C(28A)-C(16A)-C(17A)-C(22A)	0.25(18)		C(3B)-C(4B)-C(9B)-C(10B)	-5.0(2)
C(2A)-C(16A)-C(17A)-C(22A)	-119.76(17)		C(8B)-C(9B)-C(10B)-C(11B)	-4.3(4)

C(4B)-C(9B)-C(10B)-C(11B)	179.1(2)	C(22B)-C(17B)-C(18B)-C(19B)	1.2(3)
C(8B)-C(9B)-C(10B)-C(15B)	173.8(2)	C(16B)-C(17B)-C(18B)-C(19B)	-176.56(18)
C(4B)-C(9B)-C(10B)-C(15B)	-2.8(2)	C(17B)-C(18B)-C(19B)-C(20B)	-0.2(3)
C(15B)-C(10B)-C(11B)-C(12B)	-1.8(3)	C(18B)-C(19B)-C(20B)-C(21B)	-0.4(3)
C(9B)-C(10B)-C(11B)-C(12B)	176.1(2)	C(19B)-C(20B)-C(21B)-C(22B)	0.0(3)
C(10B)-C(11B)-C(12B)-C(13B)	-1.8(3)	C(20B)-C(21B)-C(22B)-C(17B)	1.0(3)
C(11B)-C(12B)-C(13B)-C(14B)	3.4(4)	C(20B)-C(21B)-C(22B)-C(23B)	178.70(19)
C(12B)-C(13B)-C(14B)-C(15B)	-1.4(3)	C(18B)-C(17B)-C(22B)-C(21B)	-1.6(3)
C(13B)-C(14B)-C(15B)-C(10B)	-2.2(3)	C(16B)-C(17B)-C(22B)-C(21B)	176.50(16)
C(13B)-C(14B)-C(15B)-C(3B)	173.00(19)	C(18B)-C(17B)-C(22B)-C(23B)	-179.77(16)
C(11B)-C(10B)-C(15B)-C(14B)	3.8(3)	C(16B)-C(17B)-C(22B)-C(23B)	-1.66(19)
C(9B)-C(10B)-C(15B)-C(14B)	-174.57(17)	C(21B)-C(22B)-C(23B)-C(24B)	2.8(3)
C(11B)-C(10B)-C(15B)-C(3B)	-172.19(17)	C(17B)-C(22B)-C(23B)-C(24B)	-179.28(19)
C(9B)-C(10B)-C(15B)-C(3B)	9.4(2)	C(21B)-C(22B)-C(23B)-C(28B)	-175.81(19)
N(1B)-C(3B)-C(15B)-C(14B)	40.3(3)	C(17B)-C(22B)-C(23B)-C(28B)	2.1(2)
C(4B)-C(3B)-C(15B)-C(14B)	172.54(19)	C(28B)-C(23B)-C(24B)-C(25B)	0.1(3)
C(2B)-C(3B)-C(15B)-C(14B)	-58.3(2)	C(22B)-C(23B)-C(24B)-C(25B)	-178.43(19)
N(1B)-C(3B)-C(15B)-C(10B)	-144.14(16)	C(23B)-C(24B)-C(25B)-C(26B)	-0.6(3)
C(4B)-C(3B)-C(15B)-C(10B)	-11.85(19)	C(24B)-C(25B)-C(26B)-C(27B)	0.3(4)
C(2B)-C(3B)-C(15B)-C(10B)	117.34(16)	C(25B)-C(26B)-C(27B)-C(28B)	0.5(3)
C(1B)-C(2B)-C(16B)-N(2B)	-79.3(2)	C(26B)-C(27B)-C(28B)-C(23B)	-1.0(3)
C(3B)-C(2B)-C(16B)-N(2B)	72.80(16)	C(26B)-C(27B)-C(28B)-C(16B)	-179.54(19)
C(1B)-C(2B)-C(16B)-C(17B)	158.81(16)	C(24B)-C(23B)-C(28B)-C(27B)	0.7(3)
C(3B)-C(2B)-C(16B)-C(17B)	50.9(2)	C(22B)-C(23B)-C(28B)-C(27B)	179.51(17)
C(1B)-C(2B)-C(16B)-C(28B)	43.8(2)	C(24B)-C(23B)-C(28B)-C(16B)	179.53(17)
C(3B)-C(2B)-C(16B)-C(28B)	-64.1(2)	C(22B)-C(23B)-C(28B)-C(16B)	-1.7(2)
N(2B)-C(16B)-C(17B)-C(18B)	-66.4(2)	N(2B)-C(16B)-C(28B)-C(27B)	62.9(2)
C(28B)-C(16B)-C(17B)-C(18B)	178.58(18)	C(17B)-C(16B)-C(28B)-C(27B)	179.34(19)
C(2B)-C(16B)-C(17B)-C(18B)	55.0(2)	C(2B)-C(16B)-C(28B)-C(27B)	-60.8(2)
N(2B)-C(16B)-C(17B)-C(22B)	115.69(16)	N(2B)-C(16B)-C(28B)-C(23B)	-115.81(16)
C(28B)-C(16B)-C(17B)-C(22B)	0.64(18)	C(17B)-C(16B)-C(28B)-C(23B)	0.65(18)
C(2B)-C(16B)-C(17B)-C(22B)	-122.94(16)	C(2B)-C(16B)-C(28B)-C(23B)	120.55(16)

8.1.6 4,4-Diphenyl-azetidin-2-one (54)

$C_{15}H_{13}NO$
orthorhombic, Pbca
a = 15.523(2) Å
b = 8.9534(9) Å
c = 17.211(3) Å
α = 90 °
β = 90 °
γ = 90 °
V = 2392.1(6) Å^3
Z = 8, $R(F)$ = 0.0893

$R_w(F^2)$ = 0.1639
Crystal size: 0.8 x 0.35 x 0.25 mm
Calculated density: 1.240 g/cm^3
2θ-Range for data collection: 5.14 – 67.45 °
Independent reflections: 2040
Observed reflections: 1020
Contributed reflections to
refinement: 2040
Refined parameters: 159

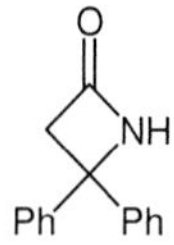

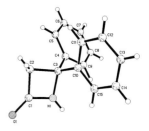

Structure

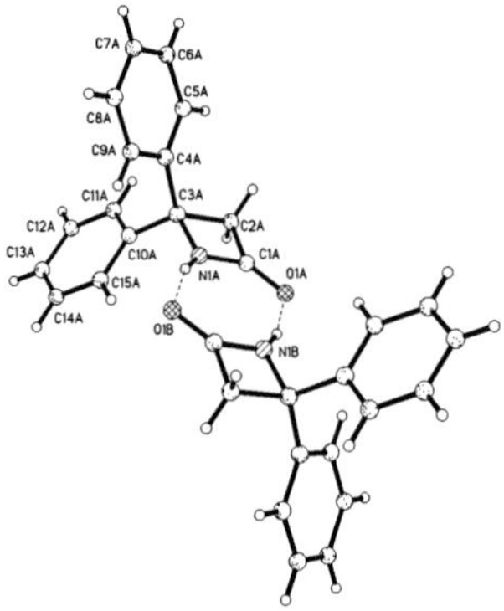

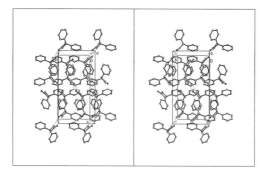

Elemental cell; View along the a-axis (a), b-axis (b) and c-axis (c):

(a)

b)

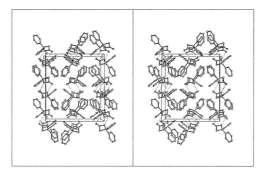

c)

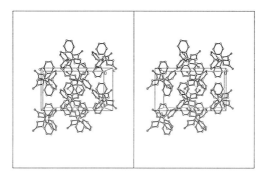

Bond lengths [Å] and angles [°]

N(1)-C(1)	1.349(6)	C(7)-H(7)	0.9300
N(1)-C(3)	1.476(6)	C(8)-C(9)	1.380(7)
N(1)-H(1)	0.87(5)	C(8)-H(8)	0.9300
O(1)-C(1)	1.221(5)	C(9)-H(9)	0.9300
C(1)-C(2)	1.521(7)	C(10)-C(15)	1.364(7)
C(2)-C(3)	1.562(6)	C(10)-C(11)	1.402(7)
C(2)-H(2A)	0.9700	C(11)-C(12)	1.368(7)
C(2)-H(2B)	0.9700	C(11)-H(11)	0.9300
C(3)-C(4)	1.492(7)	C(12)-C(13)	1.369(9)
C(3)-C(10)	1.518(6)	C(12)-H(12)	0.9300
C(4)-C(5)	1.384(7)	C(13)-C(14)	1.386(9)
C(4)-C(9)	1.388(7)	C(13)-H(13)	0.9300
C(5)-C(6)	1.404(8)	C(14)-C(15)	1.378(8)
C(5)-H(5)	0.9300	C(14)-H(14)	0.9300
C(6)-C(7)	1.354(9)	C(15)-H(15)	0.9300
C(6)-H(6)	0.9300	C(1)-N(1)-C(3)	95.7(4)
C(7)-C(8)	1.371(8)	C(1)-N(1)-H(1)	134(4)

C(3)-N(1)-H(1)	131(4)
O(1)-C(1)-N(1)	131.2(5)
O(1)-C(1)-C(2)	136.4(5)
N(1)-C(1)-C(2)	92.5(4)
C(1)-C(2)-C(3)	85.7(3)
C(1)-C(2)-H(2A)	114.4
C(3)-C(2)-H(2A)	114.4
C(1)-C(2)-H(2B)	114.4
C(3)-C(2)-H(2B)	114.4
H(2A)-C(2)-H(2B)	111.5
N(1)-C(3)-C(4)	113.5(4)
N(1)-C(3)-C(10)	112.7(4)
C(4)-C(3)-C(10)	110.0(4)
N(1)-C(3)-C(2)	86.2(3)
C(4)-C(3)-C(2)	118.0(4)
C(10)-C(3)-C(2)	114.7(4)
C(5)-C(4)-C(9)	117.4(5)
C(5)-C(4)-C(3)	122.8(4)
C(9)-C(4)-C(3)	119.7(4)
C(4)-C(5)-C(6)	119.7(5)
C(4)-C(5)-H(5)	120.1
C(6)-C(5)-H(5)	120.1
C(7)-C(6)-C(5)	121.7(6)
C(7)-C(6)-H(6)	119.1
C(5)-C(6)-H(6)	119.1
C(6)-C(7)-C(8)	119.0(6)
C(6)-C(7)-H(7)	120.5
C(8)-C(7)-H(7)	120.5
C(7)-C(8)-C(9)	120.0(6)
C(7)-C(8)-H(8)	120.0
C(9)-C(8)-H(8)	120.0
C(8)-C(9)-C(4)	122.1(5)
C(8)-C(9)-H(9)	119.0
C(4)-C(9)-H(9)	119.0
C(15)-C(10)-C(11)	117.8(5)
C(15)-C(10)-C(3)	121.7(5)
C(11)-C(10)-C(3)	120.5(5)
C(12)-C(11)-C(10)	120.1(6)
C(12)-C(11)-H(11)	119.9
C(10)-C(11)-H(11)	119.9
C(11)-C(12)-C(13)	122.1(6)
C(11)-C(12)-H(12)	119.0
C(13)-C(12)-H(12)	119.0
C(12)-C(13)-C(14)	117.6(6)
C(12)-C(13)-H(13)	121.2
C(14)-C(13)-H(13)	121.2
C(15)-C(14)-C(13)	120.8(7)
C(15)-C(14)-H(14)	119.6
C(13)-C(14)-H(14)	119.6
C(10)-C(15)-C(14)	121.5(6)
C(10)-C(15)-H(15)	119.3
C(14)-C(15)-H(15)	119.3

Torsion angles [°]

C(3)-N(1)-C(1)-O(1)	-179.6(6)
C(3)-N(1)-C(1)-C(2)	0.1(4)
O(1)-C(1)-C(2)-C(3)	179.6(6)
N(1)-C(1)-C(2)-C(3)	-0.1(4)
C(1)-N(1)-C(3)-C(4)	-119.1(4)
C(1)-N(1)-C(3)-C(10)	115.1(4)
C(1)-N(1)-C(3)-C(2)	-0.1(4)
C(1)-C(2)-C(3)-N(1)	0.1(4)
C(1)-C(2)-C(3)-C(4)	114.8(4)
C(1)-C(2)-C(3)-C(10)	-113.2(4)
N(1)-C(3)-C(4)-C(5)	130.8(5)
C(10)-C(3)-C(4)-C(5)	-101.8(6)
C(2)-C(3)-C(4)-C(5)	32.3(7)
N(1)-C(3)-C(4)-C(9)	-53.5(6)
C(10)-C(3)-C(4)-C(9)	73.8(6)
C(2)-C(3)-C(4)-C(9)	-152.1(5)
C(9)-C(4)-C(5)-C(6)	-0.2(8)
C(3)-C(4)-C(5)-C(6)	175.6(6)
C(4)-C(5)-C(6)-C(7)	0.5(10)
C(5)-C(6)-C(7)-C(8)	-1.0(10)
C(6)-C(7)-C(8)-C(9)	1.1(9)
C(7)-C(8)-C(9)-C(4)	-0.8(9)
C(5)-C(4)-C(9)-C(8)	0.3(8)
C(3)-C(4)-C(9)-C(8)	-175.5(5)
N(1)-C(3)-C(10)-C(15)	16.3(7)
C(4)-C(3)-C(10)-C(15)	-111.4(5)
C(2)-C(3)-C(10)-C(15)	112.8(5)
N(1)-C(3)-C(10)-C(11)	-163.6(4)
C(4)-C(3)-C(10)-C(11)	68.7(6)
C(2)-C(3)-C(10)-C(11)	-67.1(6)
C(15)-C(10)-C(11)-C(12)	1.9(8)
C(3)-C(10)-C(11)-C(12)	-178.2(5)
C(10)-C(11)-C(12)-C(13)	-0.2(9)
C(11)-C(12)-C(13)-C(14)	-0.9(10)
C(12)-C(13)-C(14)-C(15)	0.3(10)
C(11)-C(10)-C(15)-C(14)	-2.6(9)
C(3)-C(10)-C(15)-C(14)	177.5(6)
C(13)-C(14)-C(15)-C(10)	1.6(11)

H-Bond lengths [Å] and angles [°]

2.8941 (0.0053)	N1 - O1_$1
2.0438 (0.0528)	H1 - O1_$1
165.15 (4.84)	N1-H1-O1_$1

8.1.7 2-(9-Fluorenyl)-*N,N,N',N'*-tetramethylsuccindiamide (57)

$C_{21}H_{24}N_2O_2$
triclinic, P-1
a = 6.7215(14) Å
b = 9.3632(17) Å
c = 14.631(3) Å
α = 91.51(2)°
β = 102.954(15)°
γ = 90.119(16)°
V   = 897.0(3) Å^3

Z = 2, $R(F)$ = 0.0576
$R_w(F^2)$ = 0.1367
Crystal size: 0.8 x 0.4 x 0.3 mm
Calculated density: 1.246 g/cm^3
2θ-Range for data collection: 1.43–27.50°
Independent reflections: 4106
Observed reflections: 2817
Contributed reflections to refinement: 4106
Refined parameters: 231

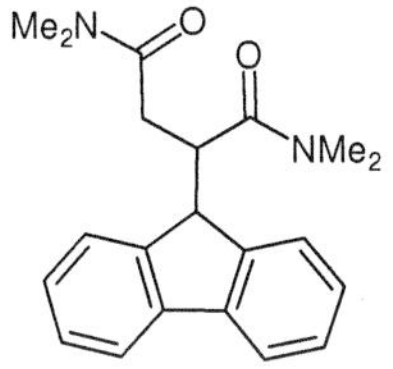

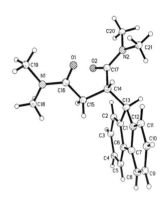

Elemental cell; View along the a-axis (a), b-axis (b) and c-axis (c):

(a)

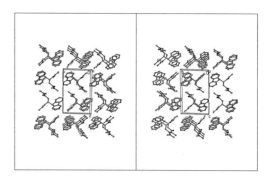

b)

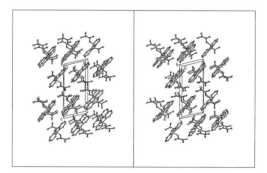

c)

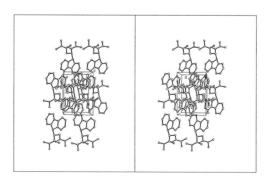

Bond lengths [Å] and angles [°]

N(1)-C(16)	1.342(3)	C(5)-C(4)-H(4)	119.7
N(1)-C(18)	1.451(3)	C(3)-C(4)-H(4)	119.7
N(1)-C(19)	1.454(3)	C(4)-C(5)-C(6)	118.9(2)
O(1)-C(16)	1.224(2)	C(4)-C(5)-H(5)	120.5
C(1)-C(2)	1.386(3)	C(6)-C(5)-H(5)	120.5
C(1)-C(6)	1.402(3)	C(5)-C(6)-C(1)	120.8(2)
C(1)-C(13)	1.516(3)	C(5)-C(6)-C(7)	130.70(19)
N(2)-C(17)	1.344(2)	C(1)-C(6)-C(7)	108.53(17)
N(2)-C(21)	1.453(3)	C(8)-C(7)-C(12)	120.4(2)
N(2)-C(20)	1.460(3)	C(8)-C(7)-C(6)	130.6(2)
O(2)-C(17)	1.224(2)	C(12)-C(7)-C(6)	109.00(17)
C(2)-C(3)	1.383(3)	C(9)-C(8)-C(7)	118.7(2)
C(2)-H(2)	0.9300	C(9)-C(8)-H(8)	120.7
C(3)-C(4)	1.385(3)	C(7)-C(8)-H(8)	120.7
C(3)-H(3)	0.9300	C(8)-C(9)-C(10)	120.9(2)
C(4)-C(5)	1.377(3)	C(8)-C(9)-H(9)	119.5
C(4)-H(4)	0.9300	C(10)-C(9)-H(9)	119.5
C(5)-C(6)	1.388(3)	C(9)-C(10)-C(11)	120.8(2)
C(5)-H(5)	0.9300	C(9)-C(10)-H(10)	119.6
C(6)-C(7)	1.461(3)	C(11)-C(10)-H(10)	119.6
C(7)-C(8)	1.391(3)	C(12)-C(11)-C(10)	118.9(2)
C(7)-C(12)	1.397(3)	C(12)-C(11)-H(11)	120.5
C(8)-C(9)	1.371(3)	C(10)-C(11)-H(11)	120.5
C(8)-H(8)	0.9300	C(11)-C(12)-C(7)	120.29(19)
C(9)-C(10)	1.377(4)	C(11)-C(12)-C(13)	129.70(19)
C(9)-H(9)	0.9300	C(7)-C(12)-C(13)	110.01(17)
C(10)-C(11)	1.390(3)	C(1)-C(13)-C(12)	102.10(15)
C(10)-H(10)	0.9300	C(1)-C(13)-C(14)	117.60(15)
C(11)-C(12)	1.370(3)	C(12)-C(13)-C(14)	112.58(16)
C(11)-H(11)	0.9300	C(1)-C(13)-H(13)	108.0
C(12)-C(13)	1.522(3)	C(12)-C(13)-H(13)	108.0
C(13)-C(14)	1.552(2)	C(14)-C(13)-H(13)	108.0
C(13)-H(13)	0.9800	C(15)-C(14)-C(17)	109.65(16)
C(14)-C(15)	1.522(3)	C(15)-C(14)-C(13)	112.85(15)
C(14)-C(17)	1.529(3)	C(17)-C(14)-C(13)	111.15(16)
C(14)-H(14)	0.9800	C(15)-C(14)-H(14)	107.7
C(15)-C(16)	1.516(3)	C(17)-C(14)-H(14)	107.7
C(15)-H(15A)	0.9700	C(13)-C(14)-H(14)	107.7
C(15)-H(15B)	0.9700	C(16)-C(15)-C(14)	109.97(15)
C(18)-H(18A)	0.9600	C(16)-C(15)-H(15A)	109.7
C(18)-H(18B)	0.9600	C(14)-C(15)-H(15A)	109.7
C(18)-H(18C)	0.9600	C(16)-C(15)-H(15B)	109.7
C(19)-H(19A)	0.9600	C(14)-C(15)-H(15B)	109.7
C(19)-H(19B)	0.9600	H(15A)-C(15)-H(15B)	108.2
C(19)-H(19C)	0.9600	O(1)-C(16)-N(1)	121.07(18)
C(20)-H(20A)	0.9600	O(1)-C(16)-C(15)	120.27(18)
C(20)-H(20B)	0.9600	N(1)-C(16)-C(15)	118.65(17)
C(20)-H(20C)	0.9600	O(2)-C(17)-N(2)	121.55(18)
C(21)-H(21A)	0.9600	O(2)-C(17)-C(14)	120.26(18)
C(21)-H(21B)	0.9600	N(2)-C(17)-C(14)	118.14(16)
C(21)-H(21C)	0.9600	N(1)-C(18)-H(18A)	109.5
		N(1)-C(18)-H(18B)	109.5
C(16)-N(1)-C(18)	124.24(18)	H(18A)-C(18)-H(18B)	109.5
C(16)-N(1)-C(19)	118.63(18)	N(1)-C(18)-H(18C)	109.5
C(18)-N(1)-C(19)	117.06(19)	H(18A)-C(18)-H(18C)	109.5
C(2)-C(1)-C(6)	119.63(19)	H(18B)-C(18)-H(18C)	109.5
C(2)-C(1)-C(13)	130.02(18)	N(1)-C(19)-H(19A)	109.5
C(6)-C(1)-C(13)	110.26(17)	N(1)-C(19)-H(19B)	109.5
C(17)-N(2)-C(21)	125.02(17)	H(19A)-C(19)-H(19B)	109.5
C(17)-N(2)-C(20)	118.73(17)	N(1)-C(19)-H(19C)	109.5
C(21)-N(2)-C(20)	115.43(17)	H(19A)-C(19)-H(19C)	109.5
C(3)-C(2)-C(1)	119.1(2)	H(19B)-C(19)-H(19C)	109.5
C(3)-C(2)-H(2)	120.4	N(2)-C(20)-H(20A)	109.5
C(1)-C(2)-H(2)	120.4	N(2)-C(20)-H(20B)	109.5
C(2)-C(3)-C(4)	121.0(2)	H(20A)-C(20)-H(20B)	109.5
C(2)-C(3)-H(3)	119.5	N(2)-C(20)-H(20C)	109.5
C(4)-C(3)-H(3)	119.5	H(20A)-C(20)-H(20C)	109.5
C(5)-C(4)-C(3)	120.5(2)	H(20B)-C(20)-H(20C)	109.5

N(2)-C(21)-H(21A)	109.5
N(2)-C(21)-H(21B)	109.5
H(21A)-C(21)-H(21B)	109.5
N(2)-C(21)-H(21C)	109.5

H(21A)-C(21)-H(21C)	109.5
H(21B)-C(21)-H(21C)	109.5

Torsion angles [°]

C(6)-C(1)-C(2)-C(3)	-0.2(3)
C(13)-C(1)-C(2)-C(3)	176.0(2)
C(1)-C(2)-C(3)-C(4)	-0.8(4)
C(2)-C(3)-C(4)-C(5)	1.1(4)
C(3)-C(4)-C(5)-C(6)	-0.4(4)
C(4)-C(5)-C(6)-C(1)	-0.5(3)
C(4)-C(5)-C(6)-C(7)	-179.0(2)
C(2)-C(1)-C(6)-C(5)	0.8(3)
C(13)-C(1)-C(6)-C(5)	-176.07(18)
C(2)-C(1)-C(6)-C(7)	179.63(18)
C(13)-C(1)-C(6)-C(7)	2.7(2)
C(5)-C(6)-C(7)-C(8)	-2.8(4)
C(1)-C(6)-C(7)-C(8)	178.6(2)
C(5)-C(6)-C(7)-C(12)	177.6(2)
C(1)-C(6)-C(7)-C(12)	-1.0(2)
C(12)-C(7)-C(8)-C(9)	0.2(3)
C(6)-C(7)-C(8)-C(9)	-179.4(2)
C(7)-C(8)-C(9)-C(10)	-0.2(4)
C(8)-C(9)-C(10)-C(11)	-0.2(4)
C(9)-C(10)-C(11)-C(12)	0.6(4)
C(10)-C(11)-C(12)-C(7)	-0.6(3)
C(10)-C(11)-C(12)-C(13)	-179.5(2)
C(8)-C(7)-C(12)-C(11)	0.2(3)
C(6)-C(7)-C(12)-C(11)	179.88(19)
C(8)-C(7)-C(12)-C(13)	179.29(18)
C(6)-C(7)-C(12)-C(13)	-1.1(2)
C(2)-C(1)-C(13)-C(12)	-179.7(2)
C(6)-C(1)-C(13)-C(12)	-3.2(2)
C(2)-C(1)-C(13)-C(14)	56.6(3)
C(6)-C(1)-C(13)-C(14)	-126.89(18)
C(11)-C(12)-C(13)-C(1)	-178.5(2)
C(7)-C(12)-C(13)-C(1)	2.5(2)
C(11)-C(12)-C(13)-C(14)	-51.5(3)
C(7)-C(12)-C(13)-C(14)	129.55(18)
C(1)-C(13)-C(14)-C(15)	47.1(2)
C(12)-C(13)-C(14)-C(15)	-71.2(2)
C(1)-C(13)-C(14)-C(17)	-76.6(2)
C(12)-C(13)-C(14)-C(17)	165.14(16)
C(17)-C(14)-C(15)-C(16)	-69.5(2)
C(13)-C(14)-C(15)-C(16)	166.05(16)
C(18)-N(1)-C(16)-O(1)	-176.3(2)
C(19)-N(1)-C(16)-O(1)	6.7(3)
C(18)-N(1)-C(16)-C(15)	4.7(3)
C(19)-N(1)-C(16)-C(15)	-172.3(2)
C(14)-C(15)-C(16)-O(1)	-22.5(3)
C(14)-C(15)-C(16)-N(1)	156.52(18)
C(21)-N(2)-C(17)-O(2)	169.3(2)
C(20)-N(2)-C(17)-O(2)	0.2(3)
C(21)-N(2)-C(17)-C(14)	-13.1(3)
C(20)-N(2)-C(17)-C(14)	177.76(17)
C(15)-C(14)-C(17)-O(2)	-29.1(2)
C(13)-C(14)-C(17)-O(2)	96.4(2)
C(15)-C(14)-C(17)-N(2)	153.36(17)
C(13)-C(14)-C(17)-N(2)	-81.2(2)

8.2 NMR spectra

a) ^{13}C NMR spectrum of (2S,3RS)-3-(1-adamantyl)-3-(N-benzylhydroxylamino)-1,2-O-isopropylidene-1,2-propanediol (**7b**), recorded in CDCl$_3$

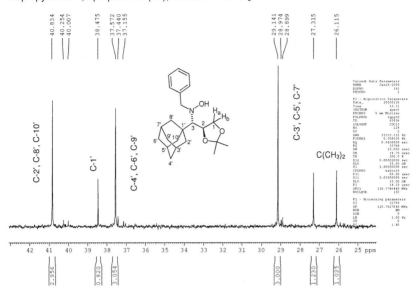

b) ^{13}C NMR spectrum of (2S,3S)-3-(1-adamantyl)-3-(tert-butoxycarbonylamino)-1,2-O-isopropylidene-1,2-propanediol (**13**), recorded in CDCl$_3$

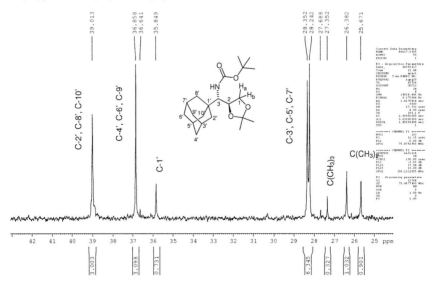

c) ^{13}C NMR spectrum of (2S,3S)-3-(1-adamantyl)-3-(N-benzylhydroxylamino)-1,2-O-cyclohexylidene-1,2-propanediol (**17**), recorded in CDCl$_3$

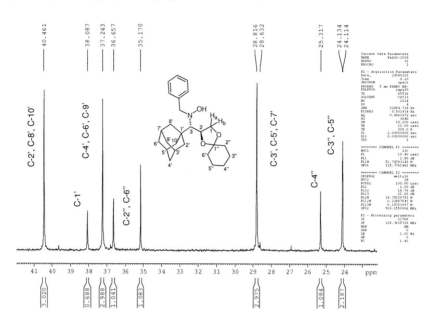

d) ^{13}C NMR spectrum of (2S,3S)-3-(1-adamantyl)-3-amino-1,2-O-cyclohexylidene-1,2-propanediol (**19**), recorded in CDCl$_3$

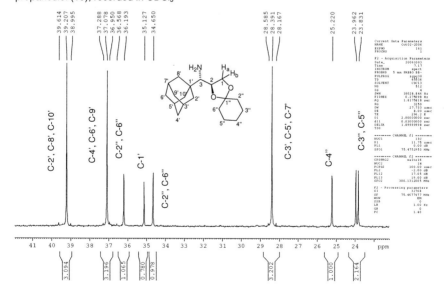

e) ^{13}C NMR spectrum of (2S,3S)-3-(1-adamantyl)-3-(tert-butoxycarbonylamino)-1,2-O-cyclohexylidene-1,2-propanediol (**20**), recorded in CDCl₃

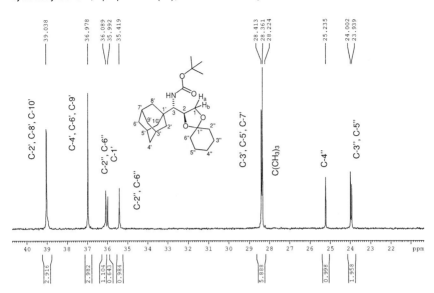

f) ^{13}C NMR spectrum of (2*S*,3*S*)-3-(1-adamantyl)-3-aminopropane-1,2-diol hydrochloride (**21**), recorded in MeOD

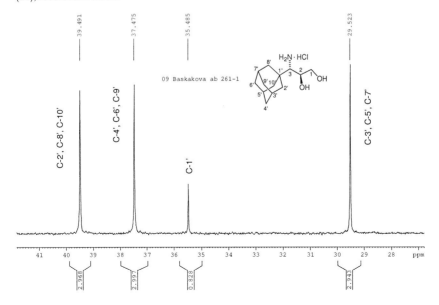

g) ^{1}H NMR spectrum of spiro[azetidine-2,9'-9H-fluorene]-4-one (**46**), recorded in CDCl$_3$

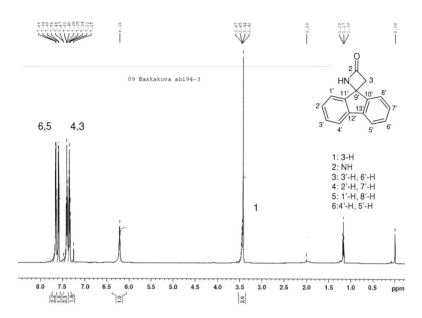

h) ^{13}C NMR spectrum of spiro[azetidine-2,9'-9H-fluorene]-4-one (**46**), recorded in CDCl$_3$

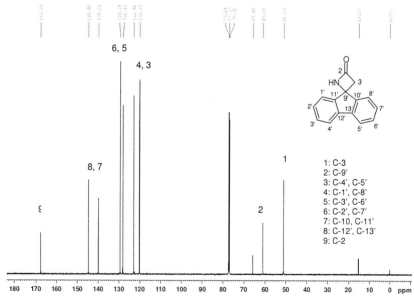

i) HSQC spectrum of spiro[azetidine-2,9'-9*H*-fluorene]-4-one (**46**)

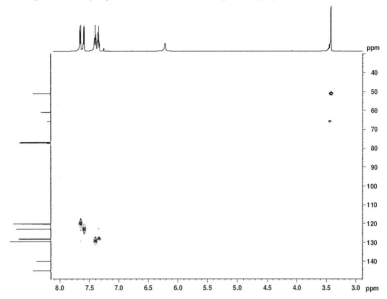

j) HMBC spectrum of spiro[azetidine-2,9'-9*H*-fluorene]-4-one (**46**)

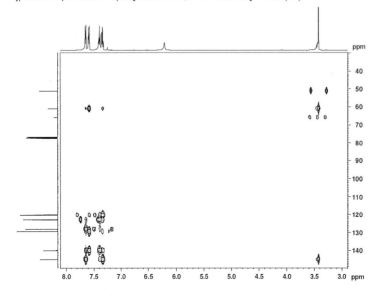

k) ¹H NMR spectrum of 3-(9-amino-9-fluorenyl)-azetidin-4-one-2-spiro-9'-fluorene (**47**), recorded in MeOD

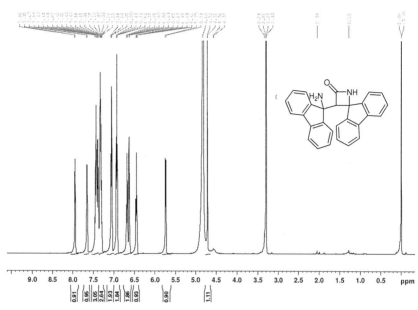

l) ¹³C NMR spectrum of 3-(9-amino-9-fluorenyl)-azetidin-4-one-2-spiro-9'-fluorene (**47**), recorded in MeOD

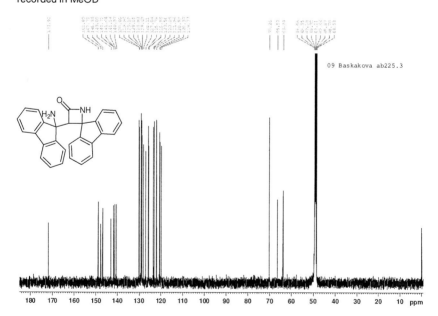

09 Baskakova ab225.3

m) HSQC spectrum of 3-(9-amino-9-fluorenyl)-azetidin-4-one-2-spiro-9'-fluorene (**47**)

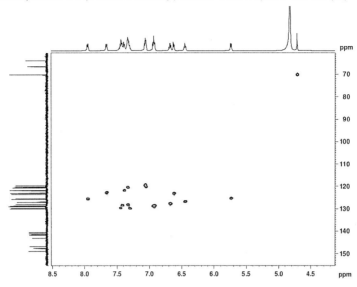

n) HMBC spectrum of 3-(9-amino-9-fluorenyl)-azetidin-4-one-2-spiro-9'-fluorene (**47**)

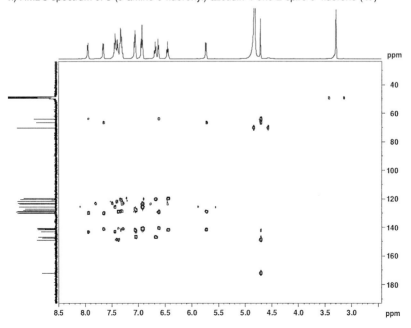

9   REFERENCES

1. http://www.rsc.org/Publishing/

2. Seebach, D.; Juaristi, E.; Miller, D. D.; Schickli, C.; Weber, T. *Helv. Chim. Acta* **1987**, *70*, 237-261.

3. Bommarius, A. S.; Schwarm, M.; Stingl, K.; Kottenhahn, M.; Huthmacher, K.; Drauz, K. *Tetrahedron: Asymmetry* **1995**, *6*, 2851-2888.

4. a) Christoffers, J.; Mann, A. *Chem. Eur. J.* **2001**, *7*, 1014-1027. – b) Christoffers, J.; Schuster, K. *Chirality* **2003**, *15*, 777-782.

5. Gálvez, N.; Moreno-Mañas, M.; Vallribera, A.; Molins, E.; Cabrero, A. *Tetrahedron Lett.* **1996**, *37*, 6197-6200.

6. Gálvez, N.; Moreno-Mañas, M.; Padros, I.; Sebastian, R. M.; Serra, N.; Vallribera, A. *Polyhedron* **1995**, *14*, 1397-1399.

7. Abele, S.; Seebach, D. *Eur. J. Org. Chem.* **2000**, 1-15.

8. Ooi, T.; Takeuchi, M.; Kameda, M.; Maruoka, K. *J. Am. Chem. Soc.* **2000**, *122*, 5228-5229.

9. Seebach, D.; Overhand M, Kühnle, F. N. M.; Martinoni, B. Oberer, L.; Hommel U.; Widmer, H. *Helv. Chimica Acta* **1996**, *79*, 913-941.

10. a) Juaristi, E.; Quintana, D.; Escalante, J. *Aldrichimica Acta* **1994**, *27*, 3-11. – b) Juaristi, E.; Quintana, D.; Lamatsch, B.; Seebach, D. *J. Org. Chem.* **1991**, *56*, 2553-2557.

11. Liu, M.; Sibi, M. P. *Tetrahedron* **2002**, *58*, 7991-8035.

12. Fülöp, F.; Martinek, T. A.; Tóth, G. K. *Chem. Soc. Rev.* **2006**, *35*, 323-334.

13. Aguilar, M.-I.; Purcell, A. W.; Devi, R.; Lew, R.; Rossjohn, J.; Smith, A. I.; Perlmutter, P. *Org. Biomol. Chem.* **2007**, *5*, 2884-2890.

14. *Elements*, *"β-Aminosäuren: Beta-Effekt befruchtet"*, Degussa Science Newletters **2004**, *6*, 8-11.

15. Minter, A.; Fuller, A. A.; Mapp, A. K. *J. Am. Chem. Soc.* **2003**, *125*, 6846-6847.

16. Liao, S.; Hruby, V. J. *Tetrahedron Lett.* **1996**, *37*, 1563-1566.

17. Georg G. I. *The Organic Chemistry of β-Lactams*, Verlag Chemie, New York, 1993.

18. Del Pozo, C.; Macías, A.; López-Ortiz, F.; Maestro, M. A.; Alonso, E.; González, J. *Eur. J. Org. Chem.* **2004**, 535-545.

19. a) Franz, T.; Hein, M.; Veith, U.; Jäger, V.; Peters, E.-M.; Peters, K.; von Schnering, H. G. *Angew. Chem.* **1994**, *106*,1308-1311; *Angew. Chem., Int. Ed. Engl.* **1994**, *33*, 1298-1301. – b) Veith, U. *Dissertation*, Universität Stuttgart, 1995.

20. Veith, U., Leurs, S., Jäger, V. *Chem. Commun.*, **1996**, 329-330.

21. Meunier, N.; Veith, U.; Jäger, V. *Chem. Commun.* **1996**, 331-332.

22. a) Schwardt, O., Veith, U., Gaspard, C., Jäger, V. *Synthesis* **1999**, 1473-1490. – b) Schwardt, O. *Dissertation*, Universität Stuttgart, 1999.

23. Bessodes, M.; Antonakis, K.; Herscovici, J.; Garcia, M.; Rochefort, H.; Capony, F.; Lelièvre, Y.; Scherman, D. *Biochem. Pharmacol.* **1999**, *58*, 329-333.

24. Hirama, M.; Hioki, H.; Itô, S. *Tetrahedron Lett.* **1988**, *29*, 3125-3128, and references cited therein.

25. Bose, A.K.; Banik, B. K.; Mathur, C.; Wagle, D. R.; Manhas, M. S. *Tetrahedron* **2000**, *56*, 5603-5619.

26. Veith, U., Schwardt, O., Jäger, V. *Synlett* **1996**, *12*, 1181-1183.

27. a) Schmid, C. R.; Bryant, J. D.; Dowlatzedah, M.; Philips, J. L.; Prather, D. E.; Schantz, R. D.; Sear, N. L.; Vianco, C. S. *J. Org. Chem.* **1991**, *56*, 4056-4058. – b) Chittenden, G. J. F. *Carbohydr. Res.* **1980**, *84*, 350-352.

28. Dumont, R.; Pfander, H. *Helv. Chim. Acta* **1983**, *66*, 814.

29. Texier-Boullet, F. *Synthesis*, **1985**, 679-681.

30. Steuer, B.; Wehner, V.; Lieberknecht, A.; Jäger, V. *Org. Synth.* **1996**, *74*, 1-12.

31. Bloch, R. *Chem. Rev.* **1998**, *98*, 1407-1438.

32. Augeri, D. J.; Robl, J. A.; Betebenner, D. A.; Magnin, D. R.; Khanna, A.; Robertson, J. G.; Wang, A.; Simpkins, L. A.; Taunk, P.; Huang, Q.; Han, S.-P.; Abboa-Offei, B.; Cap, M.; Xin,L.; Tao, L.; Tozzo, E.; Welzel, G. E.; Egan, D. M.; Marcinkeviciene, J.; Chang, S. Y.; Biller, S. A.; Kirby, M. S.; Parker, R. A.; Hamann, L. G. *J. Med. Chem.* **2005**, *48*, 5025-5037.

33. Lauster, C. D.; McKaveney, T. P.; Muench, S. V. *Am. J. Health-Syst. Pharm.* **2007**, *64*, 1265-1273.

34. Villhauer, E. B.; Brinkman, J. A.; Naderi, G. B.; Burkey, B. F.; Dunning, B. E.; Prasad, K.; Mangold, B. L.; Russell, M. E.; Hughes, T. E. *J. Med. Chem.* **2003**, *46*, 2774-2789.

35. Wanka, L.; Cabrele, C.; Vanejews, M.; Schreiner, P. R. *Eur. J. Org. Chem.* **2007**, *72*, 1474-1490.

36. Aldrich, P. E.; Hermann, E. C.; Meier, W. E.; Paulshock, M.; Prichard, W. W.; Snyder, J. A.; Watts, J. C. *J. Med. Chem.* **1971**, *14*, 535-543.

37. Rosenthal, K.; Sokol, M. S.; Ingram, R. L.; Subramanian, R.; Fort, R. C. *Antimicrob. Agents Chemother.* **1982**, *22*, 1031-1036.

38. Clariana, J., Comelles, J., Moreno-Mañas, M., Vallribera, A. *Tetrahedron: Asymmetry* **2002**, *13*, 1551-1554.

39. Belokon, Y. N.; Maleyev, V. I.; Vitt, S. V.; Ryzhov, M. G.; Kondrashov, Y. D.; Golubev, S. N.; Vauchskii, Y. P.; Kazika, A. I.; Novikova, M. I.; Krasutski, P. A.; Yurchenko, A. G.; Dubchak, I. L.; Shklover, V. E.; Struchkov, Y. T.; Bakhmutov, V. I.; Belikov, V. M. *J.*

Chem. Soc., Dalton Trans. **1985**, 17-26.

40. Clariana, J.; Garsía-Granda, S.; Gotor, V.; Gutiérrez-Fernández, A.; Luna, A.; Moreno-Mañas, M.; Vallribera, A. *Tetrahedron: Asymmetry* **2000**, *11*, 4549-4557.

41. Yoshimura, J.; Sato, T. *J. Am. Chem. Soc.* **1964**, *86*, 3858-3862.

42. Cativiela, C.; Diaz-de-Villegas, M. D.; Galvez, J. A. *Tetrahedron Lett.* **1995**, *36*, 2859-2860.

43. Badorrey, R.; Cativiela, C.; Diaz-de-Villegas, M. D.; Galvez, J. A. *Tetrahedron* **1997**, *54*, 1411-1416.

44. Badorrey, R.; Cativiela, C.; Diaz-de-Villegas, M. D.; Galvez, J. A. *Tetrahedron* **2002**, *58*, 341-354.

45. Badorrey, R.; Cativela, C.; Diaz-de-Villegas, M. D.; Diez, R.; Galvez, J. A. *Eur. J. Org. Chem.* **2003**, 2268-2275.

46. Moody, C. J.; Gallagher, P.T.; Lightfoot, A. P.; Slawin, A. M. Z. *J. Org. Chem.* **1999**, *64*, 4419-4425.

47. Madhan, A.; Kumar, A. R.; Rao, B. V. *Tetrahedron: Asymmetry* **2001**, *12*, 2009-2011.

48. Madhan, A.; Rao, B. V. *Tetrahedron Lett.* **2003**, *44*, 5641-5643.

49. Matsumoto, T.; Kobayashi, Y.; Takemoto, Y.; Ito, Y.; Kamijo, T.; Harada, H.; Terashima, S. *Tetrahedron Lett.* **1990**, *31*, 4175-4176.

50. Schnabel, G. *Dissertation*, Universität Würzburg, 1990.

51. Merino, P.; Castillo, E.; Merchan, F. L.; Tejero, T. *Tetrahedron: Asymmetry* **1997**, *8*, 1725-1729.

52. Merino, P.; Anoro, S.; Castillo, E.; Merchan, F.; Tejero, T. *Tetrahedron: Asymmetry* **1996**, *7*, 1887-1890.

53. Merino, P.; Castillo, E.; Franco, S.; Merchan, F. L.; Tejero, T. *Tetrahedron* **1998**, *54*, 12301-12322.

54. Dondoni, A.; Franco, S.; Merchan, F. L.; Merino, P.; Tejero, T. *Tetrahedron Lett.* **1993**, *34*, 5475-5478.

55. Merino, P.; Castillo E.; Franco, S.; Merchan, F. L.; Tejero, T. *J. Org. Chem.* **1998**, *63*, 2371-2374.

56. Dondoni, A. *Synth. Commun.* **1994**, *24*, 2537-2550.

57. Liu, K.-C.; Shelton, B. R.; Howe, R. K. *J. Org. Chem.* **1980**, *45*, 3916-3918.

58. Maskill, H.; Jencks, W. P. *J. Am. Chem. Soc.* **1987**, *109*, 2062-2070.

59. Borch, R. F.; Bernstein, M. D., Durst, H. D. *J. Am. Chem. Soc.* **1971**, *93*, 2897-2904.

60. Poch, M.; Alcón, M.; Moyano, A.; Pericàs, Riera, A. *Tetrahedron Lett.* **1993**, *34*, 7781-7784.

61. a) Zimmermann, P. J.; Blanarikova, I.; Jäger, V. *Angew. Chem.* **2000**, *112*, 936-938;

Angew. Chem. Int. Ed. Engl. **2000**, *39*, 910-912. – b) Zimmermann, P. J.; Lee, J.-Y.; Blanarikova, I.; Endermann, R.; Häbich, D.; Jäger, V. *Eur. J. Org. Chem.* **2005**, 3450-3460.

62. Lee, J.-Y.; Schiffer, G.; Jäger, V. *Org. Lett.* **2005**, *7*, 2317-2320.

63. Brückner, R. *Reaktionsmechanismen*; Heidelberg ; München : Elsevier, Spektrum Akad.-Verl., 2007.

64. a) Fluka Chemical Catalog, 2007-2008, p. 329. b) Pearson, A. J.; Chelliah, M. . *J. Org. Chem.* **1998**, *63*, 3087-3098.

65. Marcías, A., Alonso, E., del Pozo, C., Venturini, A., Gonzáles, J. *J. Org. Chem.* **2004**, *69*, 7004-7012.

66. List, B., Pojarliev, P., Biller, W. T., Martin, H. J. *J. Am. Chem. Soc.* **2002**, *124*, 827-833.

67. Chattopadhyay, A. *J. Org. Chem.* **1996**, *61*, 6104-6107.

68. Majewski, M.; Nowak, P. *J. Org. Chem.* **2000**, *65*, 5152-5160.

69. Kalinowski, H.-O.; Berger, S.; Braun, S. "^{13}C NMR Spektroskopie", Georg Thieme Verlag Stuttgart, New York **1984**.

70. Juaristi, E. „Introduction to Stereochemistry and Conformational Analysis", A Wiley-Interscience Publication, John Wiley & Sons, Inc.

71. Wenzel, A.; Jacobsen, E. N. *J. Am. Chem. Soc.* **2002**, *124*, 12964-12965.

72. Jacobsen, M. F.; Ionita, I.; Skrydstrup, T. *J. Org. Chem.* **2004**, *69*, 4792-4796.

73. Henneböhle, M.; Le Roy, P.-Y.; Hein, M.; Ehrler, R.; Jäger, V. *Z. Naturforsch., B: Chem. Sci.* **2004**, *59*, 451-467.

74. Angelaud, R.; Zhong, Y.-L.; Maligres, P.; Lee, J.; Askin, D. *J. Org. Chem.* **2005**, *70*, 1949-1952.

75. Josephsohn, N. S.; Carswell, E. L.; Snapper, M. L.; Hoveyda, A. H. *Org. Lett.* **2005**, *7*, 2711-2713.

76. Deboves, H. J. C.; Grabowska, U.; Rizzo, A., Jackson, R. F. W. *J. Chem. Soc., Perkin Trans.* 1, **2000**, 4284-4292.

77. Liu et al.; Cardillo, G.; Tomasini, C. *Chem. Soc. Rev.* **1996**, 117-128.

78. Hamed, O.; Henry, P. M. *Organometallics* **1997**, *16*, 4903-4909.

79. Fujisawa, H.; Takahashi, E.; Mukaiyama, T. *Chem. Eur. J.* **2006**, *12*, 5082-5093.

80. Palomo, C.; Aizpurua, J. M.; Ganboa, I.; Oiarbide, M. *Synlett.* **2001**, *12*, 1813-1826.

81. Bolli, M. H.; Marfurt, J.; Grisostomi, C.; Boss, C.; Binkert, C.; Hess, P.; Treiber, A.; Thorin, E.; Morrison, K.; Buchmann, S.; Bur, D.; Ramuz, H.; Clozel, M.; Fischli, M.; Weller, T. *J. Med. Chem.* **2004**, *47*, 2776-2795.

82. Gerona-Navarro, G.; García-López, M. T.; González-Muñiz, R. *Tetrahedron Lett.* **2003**, *44*, 6145-6148.

83. Angelaud, R.; Zhong, Y.-L.; Maligres, P.; Lee, J.; Askin, D. *J. Org. Chem.* **2005**, *70*, 1949-1952.

84. Podlech, J.; Seebach, D. *Liebigs Ann.* **1995**, 1217-1228.

85. Cainelli, G.; Giacomini, D.; Mezzina, E.; Panunzio, M.; Zarantonello, P. *Tetrahedron Lett.* **1991**, *32*, 2967-2970.

86. Cainelli, G.; Giacomini, D.; Panunzio, M.; Zarantonello, P. *Tetrahedron Lett.* **1992**, 7783-7786.

87. Gyenes, F.; Bergmann, K. E.; Welch, J. T. *J. Org. Chem.* **1998**, *63*, 2824-2828.

88. Cainelli, G.; Panunzio, M.; Andreoli, P.; Martelli, G.; Spunta, G.; Giacomini, D.; Bandini, E. *Pure Appl. Chem.* **1990**, *62*, 605-612, and references sited herein.

89. Yamamoto, Y.; Komatsu, T.; Maruyama, K. *J. Am. Chem. Soc.* **1984**, *106*, 5031-5033.

90. Yamamoto, Y.; Nishii, S.; Maruyama, K.; Komatsu, T.; Ito, W. *J. Am. Chem. Soc.* **1986**, *108*, 7778-7786.

91. Itsuno, S.; Watanabe, K.; Ito, K.; El-Shehawy, A. A.; Sarhan, A. A. *Angew. Chem.* **1997**, *109*, 105-107.

92. Ramachandran, P. V. *Aldrichimica Acta* **2002**, *35*, 23-35.

93. Gung, B. W. *Org. Reactions*, *64*, Edited by Larry E. Overman et al.

94. Brown, H. C.; Chandrasekharan, J.; Ramachandran, P. V. *J. Am. Chem. Soc.* **1988**, *110*, 1539-1546.

95. Armesto, D.; Ortiz , M. J.; Perez-Ossorio, R. *J. Chem. Soc. Perkin Trans. I* **1986**, 2021-2026.

96. Schomaker, J. M.; Travis, B. R.; Borhan, B. *Org. Lett.* **2003**, *5*, 3089-3092.

97. Gobbi, S.; Rampa, A.; Bisi, A.; Belluti, F.; Piazzi, L.; Valenti, P.; Caputo, A., Zampiron, A.; Carrara, M. *J. Med. Chem.* **2003**, *46*, 3662-3669.

98. Minter, A. R.; Brennan, B. B.; Mapp, A. K. *J. Org. Chem.* **2004**, *126*, 10504-10505.

99. Norrby, P.-O.; Becker, H.; Sharpless, K. B. *J. Am. Chem. Soc.* **1996**, *118*, 35-42.

100. Schwardt, O. *Dissertation*, Universität Stuttgart, 1999.

101. Smith, A. B., III; Leahy, J. W.; Noda, I.; Remiszewski, S. W.; Lirerton, N. J.; Zibuck, R. *J. Am. Chem. Soc.* **1992**, *114*, 2995-3007.

102. Adams, H.; Anderson, J. C.; Peace, S.; Pennell, A. M. K. *J. Org. Chem.* **1998**, *63*, 9932-9934.

103. Zimmerman, P. *Dissertation*, Universität Stuttgart, 2000.

104. Levine, A. W. , Fech, J. *J. Org. Chem.* **1972**, *37*, 1500-1503.

105. Ritter, J. J.; Kalish, J. *J. Am. Chem. Soc.***1948**, *70*, 4048-4050.

106. Laxma Reddy, K. *Tetrahedron Letters*, **2003**, *44*, 1453-1455.

107. a) Itazaki, M.; Nishihara, Y.; Osakada, K. *J. Org. Chem.* **2002**, *67*, 6889-6895. – b) Shi,

M.; Wang, B.-Y.; Huang, J.-W. *J. Org. Chem.* **2005**, *70*, 5606-5610.

108. Ritter, J. J.; Kalish, J. *Org. Synth.*, **1973**, *Coll. Vol. 5*, 471, **1964**, *44*, 44.

109. DuPriest, M. T.; Conrow, R. E.; Kuzmich, D. *Tetrahedron Lett.* **1990**, *31*, 2959-2962.

110. Norrby, P.-O.; Becker, H.; Sharpless, K. B. *J. Am. Chem. Soc.* **1996**, *118*, 35-42.

111. Hart, D. J.; Kanai, K.-i.; Thomas, D.G.; Yang, T.-K. *J. Org. Chem.* **1983**, *48*, 289-294.

112. a) Mikami, K.; Matsumoto, S.; Ispida, A.; Takamuku, S.; Suenobu, T.; Fukuzumi, S. *J. Am. Chem. Soc.* **1995**, *117*, 11134-11141. – b) Zou, B.; Wei, J.; Cai, G.; Ma, D. *Org. Lett.* **2003**, *5*, 3503-3506.

113. Krüger, C.; Rochow, E. G.; Wannagat, U. *Chem. Ber.* **1963**, *96*, 2132-2137.

114. Basel, Y.; Hassner, A. *J. Org. Chem.* **2000**, *65*, 6368-6380.

115. Del Río, E.; López, R.; Menéndez, M. I.; Sordo, T. L.; Ruiz-López, M. F. J. Comput. Chem. **1998**, *19*, 1826-1833.

116. Ha, D.-C.; Hart, D. J.; Yang, T.-K. *J. Am. Chem. Soc.* **1984**, *106*, 4819-4825.

117. Gluchowski, C.; Cooper, L.; Bergbreiter, D. E.; Newcomb, M. *J. Org. Chem.* **1980**, *45*, 3413-3416.

118. Fujisawa, T.; Ukaji, Y.; Noro, T.; Date, K.; Shimizu, M. *Tetrahedron Lett.* **1991**, *32*, 7563-7566.

119. Shimizu, M.; Ukaji, Y.; Tanizaki, J.; Fujisawa, T. *Chem. Lett.* **1992**, 1349-1352.

120. Hart, D. J.; Kanai, K.-i.; Thomas, D.G.; Yang, T.-K. *J. Org. Chem.* **1983**, *48*, 289-294.

121. a) Zou, B., Wei, J., Cai, G., Ma, D. *Org. Lett.* **2003**, *5*, 3503-3506. – b) Rubio, A., Liebeskind, L. S. *J. Am. Chem. Soc.* **1993**, *115*, 891-901.

122. Courtney, M. C.; MacCormack, A. C.; More O'Ferrall, R. A. *J. Phys. Org. Chem.* **2002**, *15*, 529-539.

123. Staudinger, H. *Liebigs Ann. Chem.* **1907**, *356*, 51-123.

124. a) Krämer, B.; Franz, T.; Picasso, S.; Pruschek, P.; Jäger, V. *Synlett.* **1997**, 295-297. – b) Palomo, C.; Aizpurua, J. M.; García, J. M.; Galarza, R.; Legido, M.; Urchegui, R.; Román, P.; Luque, A.; Server-Carrió, J.; Linden, A. *J. Org. Chem.* **1997**, *62*, 2070-2079.

125. Griesser, H.; Vogt, S.; Jäger, V. *Unpublished results*, Universität Stuttgart, 2005-2007.

126. a) Ballestri, M.; Chatgilialoglu, C. *J. Org. Chem.* **1991**, *56*, 678-683. – b) Chatgilialoglu, C. *Acc. Chem. Res.* **1992**, *25*, 188-194.

127. Moriconi, E. J.; Kelly, J. F.; Salomone, R. A. *J. Org. Chem.* **1968**, *33*, 3448-3452.

128. Verkade J. M. M.; van Hemert, L. J. C.; Quaedflieg, P. J. L. M.; Alsters, P. L.; van Delft, F. L.; Rutjes, F. P. J. T. *Tetrahedron Lett.* **2006**, *47*, 8109-8113.

129. Günter, H. *NMR Spectroscopy* John Wiley & Sons Inc., 2nd Edition, 1994.

130. Fujisawa, T.; Ukaji, Y.; Noro, T.; Date, K.; Shimizu, M. *Tetrahedron* **1992**, *48*, 5629-5638.

131. Mukaiyama, T.; Goto, Y.; Shoda, S.-i. *Chem. Lett.* **1983**, 671-674.

132. Woodbury, R. P.; Rathke, M. W. *J. Org. Chem.* **1977**, *42*, 1688-1690.

133. Greene, T. W.; Wuts, P. G. M. *Protective Groups in Organic Synthesis* John Wiley & Sons, Inc, Second Edition, 1991.

134. Gassman, P. G.; Hodgson, P. K. G.; Balchunis, R. J. *J. Am. Chem. Soc.* **1976**, *98*, 1275-1276.

135. a) Enders, D.; Kirchhoff, J.; Gerdes, P.; Mannes, D.; Raabe, G.; Runsink, J.; Boche, G.; Marsch, M.; Ahlbrecht, H.; Sommer, H. *Eur. J. Org. Chem.* **1998**, 63-72. – b) Evans, G. B.; Furneaux, R. H.; Hausler, H.; Larsen, J. S.; Tyler, P. C. *J. Org. Chem.* **2004**, *69*, 2217-2220.

136. Minter, A. R.; Brennan, B. B.; Mapp, A. K. *J. Org. Chem.* **2004**, *126*, 10504-10505.

137. Flügge, J.*Grundlagen der Polarimetrie*, de Gruyter-Verlag, Berlin, **1970**, 16.

138. Sheldrick, G. *Program SHELXS-86 und SHELXL-93*, Institut für Anorganische Chemie der Universität Göttingen, **1986**, **1993**.

139. Stewart, J. M.; Dickinson, P. A.; Ammon, H. L.; Flach, H.; Heck, H. *Programm XRAY-76*, Tech. Rep. TR-446, University of Maryland, Computer Center, College Park MD, **1976**.

140. Johnson, C. K. Programm ORTEP II, Tech. Rep. ORNL-5138, Oak Ridge National Laboratory, Oak Ridge, TN, **1971**.

141. Hildenbrand, T.; *Programm FRIEDA*, Universität Stuttgart, unpublished.

142. Jork, H.; Funk, W.; Fischer, W.; Wimmer, H. *Dünnschicht-Chromatographie, Reagenzien und Nachweismethoden*, Bd. 1a, VCH, Winheim, **1989**.

143. Helmchen, G.; Glatz, B. Ein apparativ einfaches System und Säulen höchster Trenn-Leistung zur präparativen Mitteldruckchromatographie, Anhang zur Habilitationsschrift, Stuttgart, **1978**.

144. Armarego, W. L. F.; Perrin, D. D. *Purification of Laboratory Chemicals*, 4th Edition, Butterworth Heinemann, **2002**.

145. U.Veith *Diplomarbeit*, Universität Würzburg, 1991.

146 .Wang, M. X.; Lin, Sh.-J. *J. Org. Chem.* **2002**, *67*, 6542-6545.

147 .Menzel, A. *Dissertation*, Universität Stuttgart, 2000.

148. Canas-Rodriguez, A.; Ruiz-Poveda, S. G.; Coronel-Borges, L. A. *Carbohydr. Res.* **1987**, *159*, 217-227.

149. Hasegawa, M.; Taniyama, D.; Tomioka, K. *Tetrahedron* **2000**, *56*, 10153-10158.

150. *Diazald® and Diazomethane Generators*,
     http://www.sigmaaldrich.com/etc/medialib/docs/Aldrich/Bulletin/al_techbull_al180.Par.0
     001.File.tmp/al_techbull_al180.pdf

151. Kohmura, Y., Mase, T. *J. Org. Chem.* **2004**, *69*, 6329-6334.

152. Hudson, R. F., Brown, C., Maron, A. *Chem. Ber.* **1982**, *115*, 2560-2573.

153. Reddy, K. Laxma *Tetrahedron Letters*, **2003**, *44*, 1453-1455.

154. Jiang, W.-Q.; Costa, S. P. G.; Maia, H. L. S. *Org. Biomol. Chem.* **2003**, *1*, 3804-3810.

10 ACKNOWLEDGEMENTS

Having had the opportunity to work and study in Jäger's group for the last five years has been a privilege, and even, at times, a pleasure. I would like to think that I have grown as a person and as a scientist in this time. I thank Professor Volker Jäger for letting me work in his group and for the continuous supervision of my Thesis.

I am very grateful to Frau Gisela Kraschewski-Fien for her help in all bureaucratic problems, her enthusiasm and willingness to organize some activities outside of work, and to Herrn Dipl.-Ing. Helmut Griesser for his competent technical mentoring and fruitful tips.

Working in the Jäger's group has been a precious experience. I want to thank the "old school" – Ja Young Lee, Jörg Williardt, Sunitha Shiva, Robert Sardzik, and of course my present colleagues from various nations: Jeanne, Christof, Amélie, Sony, Mohammad, Hua (in no particular order), who made the working atmosphere warm and vivid. Finally, I would like to thank Mrs. Dr. Hend El Sehrawi, a guest scientist from Cairo/Egypt, for her kind cooperation in one of my projects.

Many thanks go to all the employees of the spectroscopical and analytical departments for the careful measurements and also to Dr. P. Fischer and Dr. B. Klaasen for their help with the NMR spectra. This PhD thesis would not be complete without invaluable efforts of our department crystallographer Dr. Wolfgang Frey on the X-ray analyses. I must also acknowledge Dr. Burkhard Miehlich for his "rescue service" for my computer and installations of programs.

I am also thankful to my research students Friedrich Wartlick, Andreas Bogner and Michael Krebs, for their helpful experimental contributions to this work.

My research project has been financially supported by Altana Pharma AG, and I am personally grateful to Prof. Dr. Jörg Senn-Bilfinger.

I want to thank my parents for being a solid backup, and the part of my family here in Germany (especially my lovely aunt), who have always been next to me when I was in need and helped not to forget my native language.

My deepest thanks go to Jules, who in last three years has been, among many other things, a cheerful person, a brilliant partner for scientific discussions, and a source of dear love and encouragement. This Thesis and all of the work that went into it would not have been possible without him.

And finally, last but certainly not least, I want to acknowledge José, Fadi, Pegor, Talal, Ora and Pedro for the amusing holidays, trips and weekends.

11 CURRICULUM VITAE

Personal Information

Name	Baskakova, Alevtina
Date and place of birth	27.06.1981 in Emwa, Republic of Komi, Russian Federation
Nationality	Russian
Marital status	Single

Schools

1988-1996	Elementary and Secondary Schools / Voronezh and Emwa, Russia
1996-1998	High School / Saint Petersburg, Russia
June 1998	High School Examination Certificate

Academic Qualifications

Sept. 1998-June 2003	Undergraduate Studies at the Faculty of Chemistry, St. Petersburg State University, St. Petersburg, Russia
June 2003	Graduated with a degree equivalent to M.Sc. of Chemistry with Diploma Thesis: "Synthesis of Estrogen Analogs Substituted in B- and/or D-Rings", supervisor Prof. Dr. A. G. Shavva, Department of Natural Products Chemistry, Faculty of Chemistry, St. Petersburg State University, St. Petersburg
July 2003 – May 2004	Scientific and industrial projects in the group of Prof. Dr. A. de Meijere, Institut für Organische Chemie, Georg-August Universität, Göttingen, Germany
Since June 2004	Conferral of a doctorate in the group of Prof. Dr. V. Jäger, Institut für Organische Chemie, Universität Stuttgart, Germany

Employments

July 2003 – May 2004	Research assistent in the company KAdemCustomChem, Institut für Organische Chemie, Georg-August Universität, Göttingen
July 2004 – Sept. 2004	Wissenschaftliche Hilfskraft am Institut für Organische Chemie, Universität Stuttgart
Since Okt. 2004	Wissenschaftliche Angestellte am Institut für Organische Chemie, Universität Stuttgart

12 FORMULA TABLE OF COMPOUNDS PREPARED

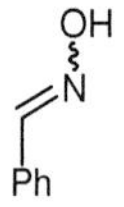

1

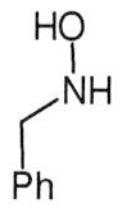

2

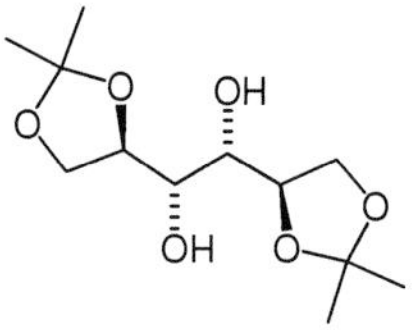

3

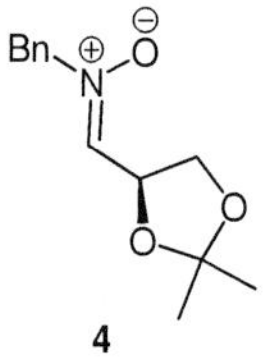

4

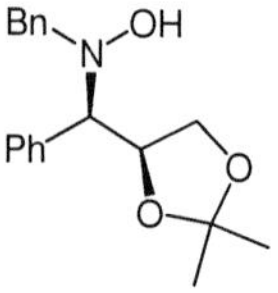

5a

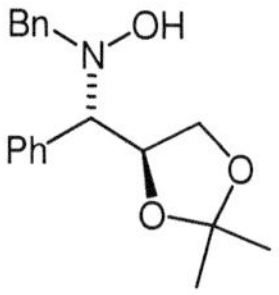

5b

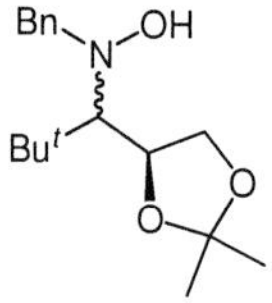

6a/b

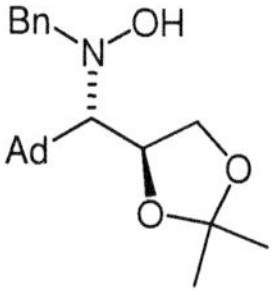

7b

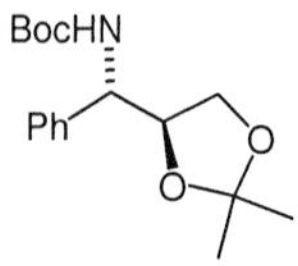

8a

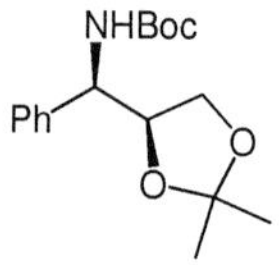

8b

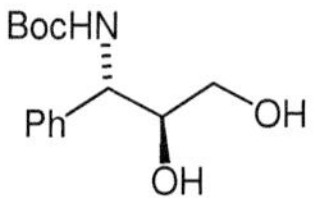

9

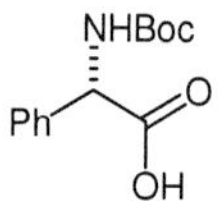

10

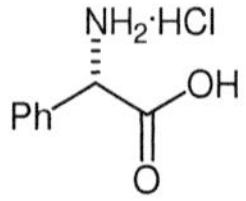

(S)-11·HCl

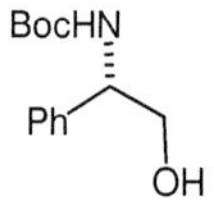

12

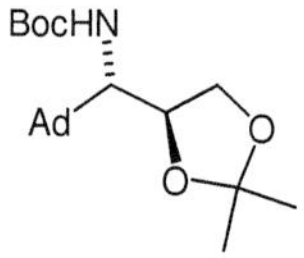

13

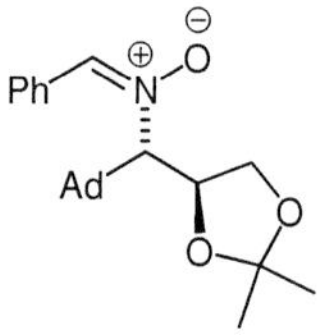

14

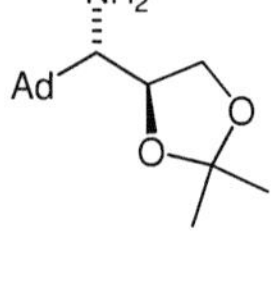

15

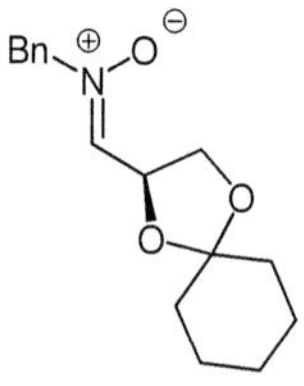

16

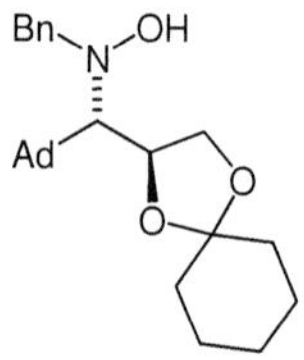

17

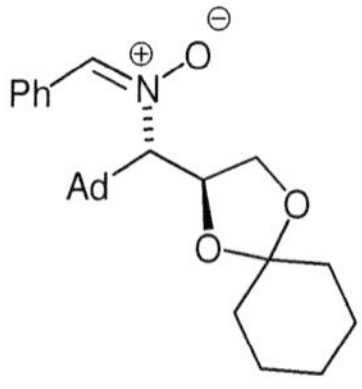

18

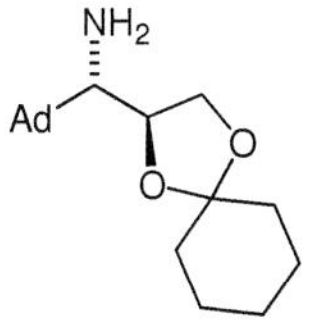

19

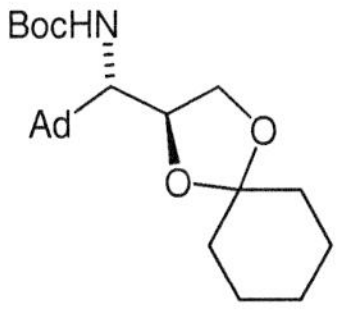

20

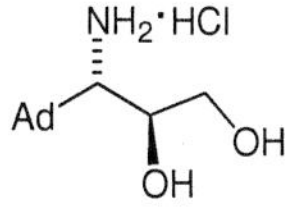

21

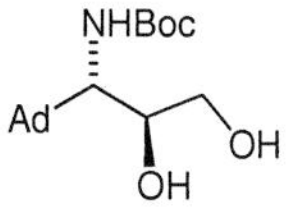

22

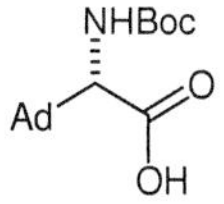

(S)-**23**

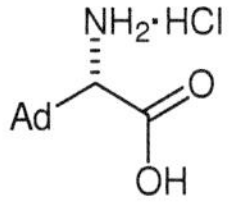

(S)-**24**

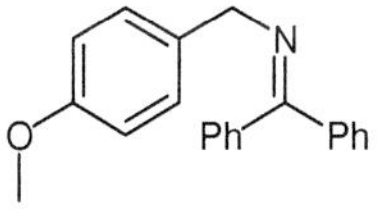

25

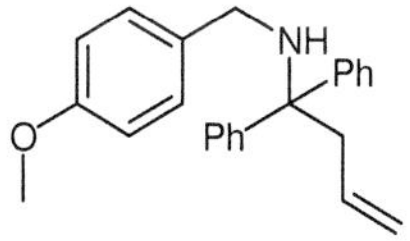

26

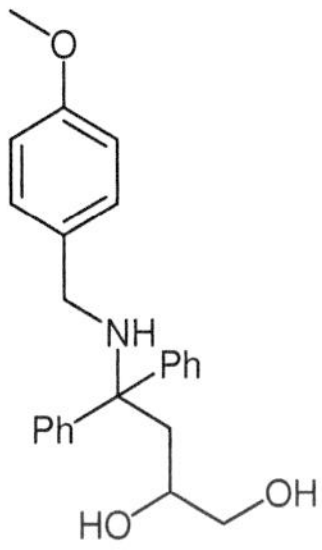

27

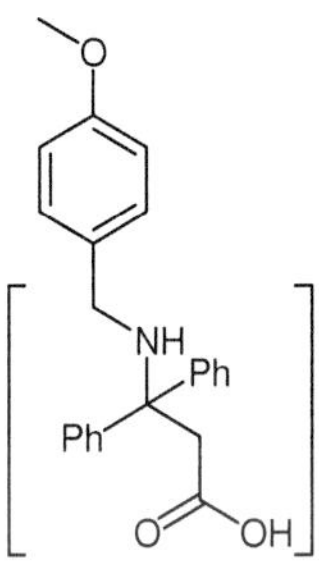

28

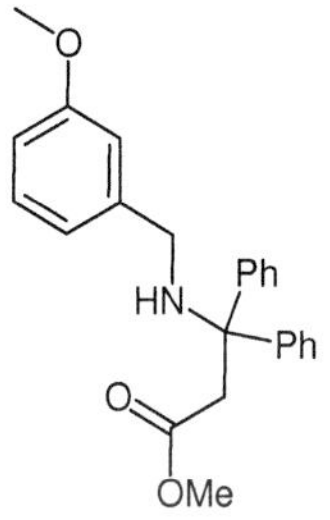

29

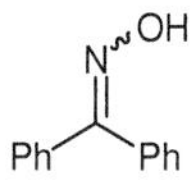

30

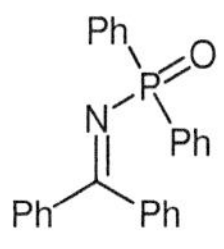

31

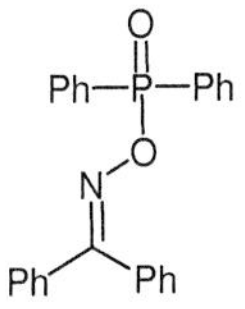

32

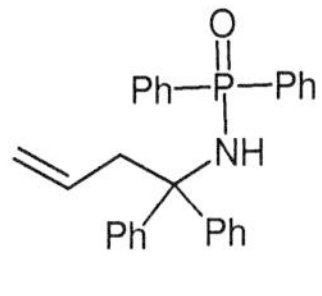

33

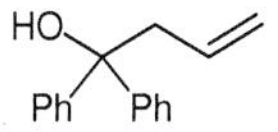

34

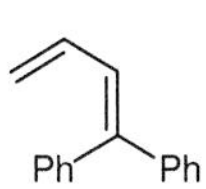

35

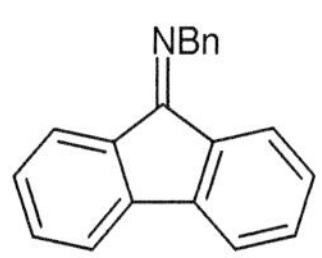

36

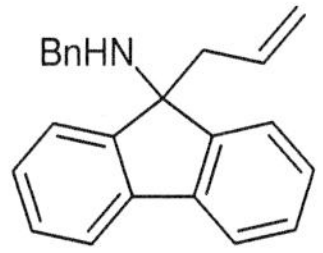

37

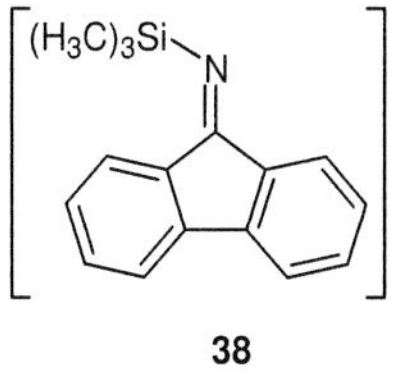

38

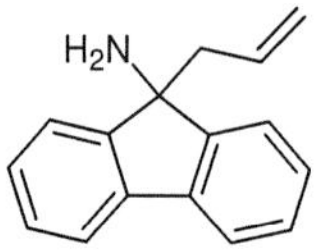

39

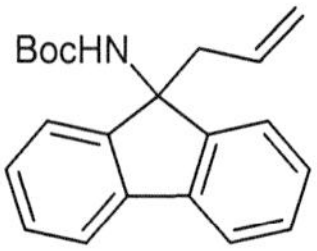

40

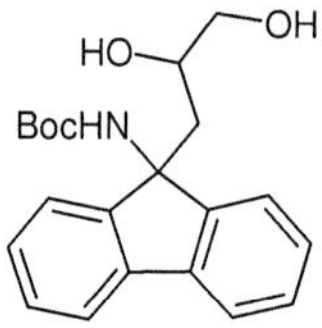

41

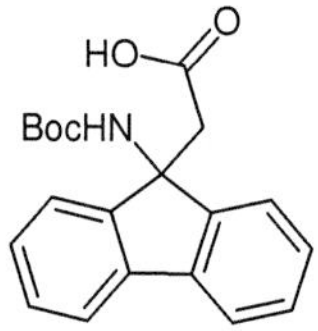

42

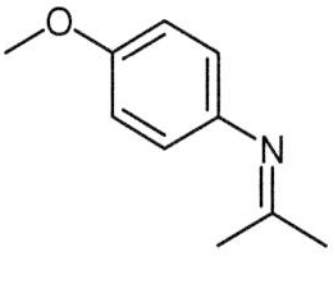

43

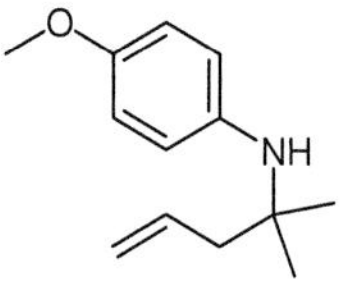

44

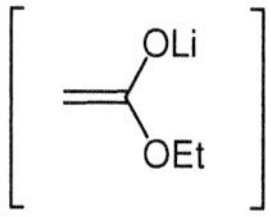

45

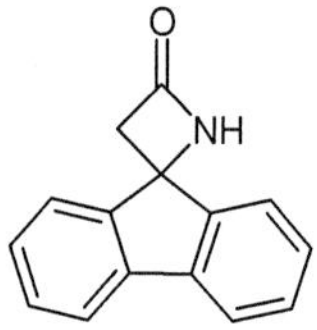

46

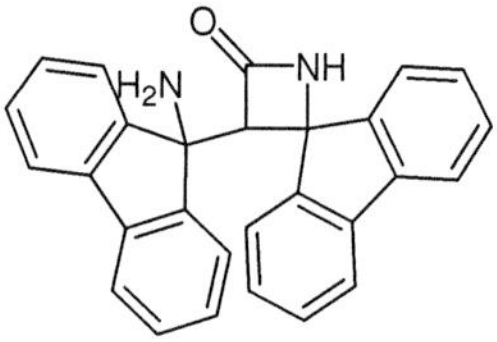

47

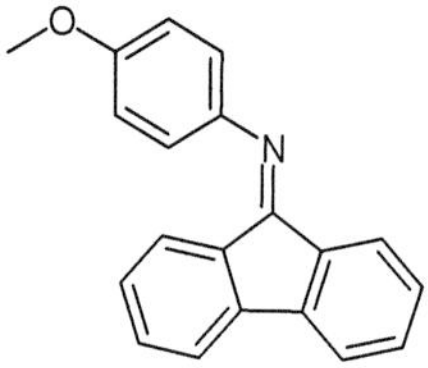

48

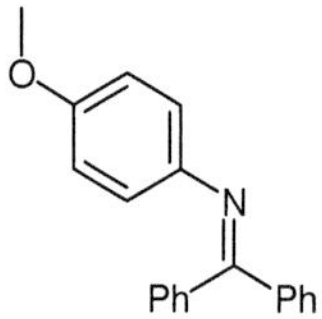

49

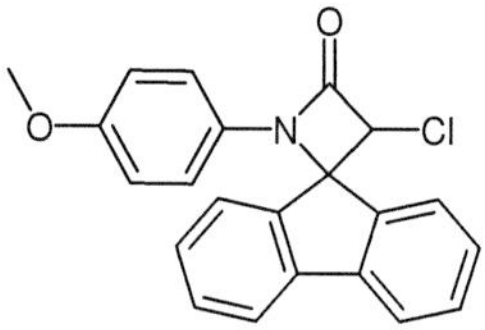

50

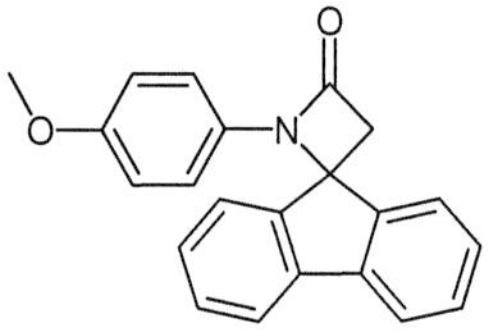

51

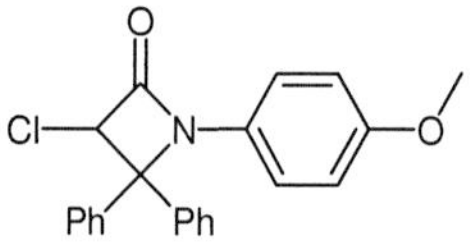

52

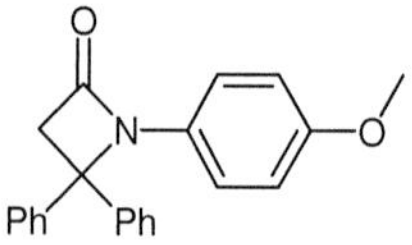

53

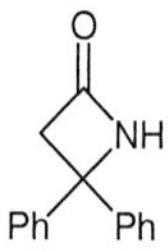

54

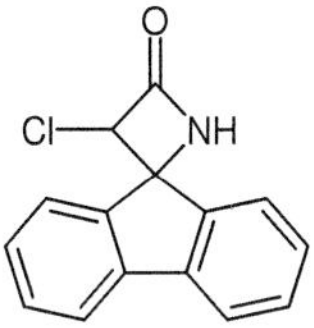

55

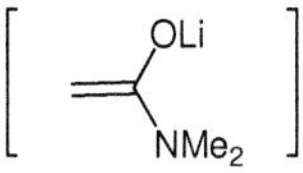

56

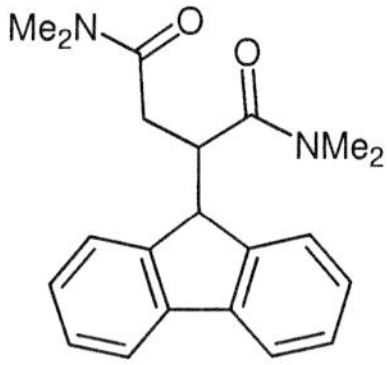

57

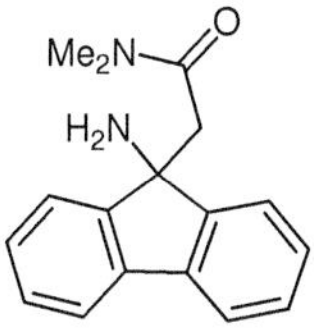

58

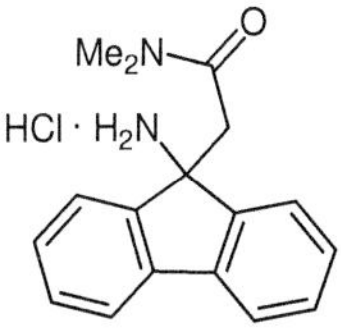

59

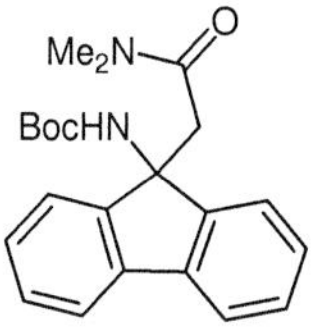

60